COMPACTION OF SOILS, GRANULATES AND F

T0231289

ADVANCES IN GEOTECHNICAL ENGINEERING AND TUNNELLING

3

General editor:

D. KOLYMBAS

University of Innsbruck, Institute of Geotechnics and Tunnelling

INTERNATIONAL WORKSHOP ON COMPACTION OF SOILS, GRANULATES AND POWDERS / INNSBRUCK / 28-29 FEBRUARY 2000

Compaction of Soils, Granulates and Powders

Edited by

D. KOLYMBAS & W. FELLIN

University of Innsbruck, Institute of Geotechnics and Tunnelling

A.A. BALKEMA / ROTTERDAM / BROOKFIELD / 2000

Published by
A.A. Balkema, P.O. Box 1675, 3000 BR Rotterdam, Netherlands
Fax: +31.10.4135947; E-mail: balkema@balkema.nl; Internet site: http://www.balkema.nl

A.A. Balkema Publishers, Old Post Road, Brookfield, VT 05036-9704, USA
Fax: 802.276.3837; E-mail: info@ashgate.com

ISSN 1566-6182
ISBN 90 5809 317 4 hardbound edition
ISBN 90 5809 318 2 student paper edition

Table of contents

3 On-line compaction control

4 Quasi static experimental investigations, cyclic loading

5 Dynamic experimental investigations

6 Theoretical approaches

7 Numerical investigations: Distinct element models, lattice models

8 Numerical investigations: Continuum models

Late paper

Preface

Granular materials exhibit a peculiar and fascinating behaviour and are the subject of intensive research. In particular, the way granular materials can be compressed (or compacted) attracts much interest not only from a theoretical but also from a practical point of view since e.g. a dense sand is an excellent foundation ground whereas a loose sand is a very dangerous one. Unlike other particulate materials, such as gases, the compaction of granulates is not a function of the applied pressure. Experience shows that cyclic and/or dynamic loading considerably enhances compaction. The underlying mechanism of dynamic compaction are not yet completely understood. Things become even more complicate if we consider water-saturated granulates. The entrapped pore-water impedes compaction if the compacting stress is applied in a quasi-static way. For dynamic loading, however, pore-water seems to facilitate compaction in many cases! The proceedings of the Workshop on Compaction of Soils, Granulates and Powders organised by the Institute of Geotechnics and Tunnelling of the University of Innsbruck, 28 - 29 February 2000 are devoted to these problems.

D. Kolymbas, W. Fellin

1 Practical applications in engineering

Newest innovations into soil and asphalt compaction technology

Rudolf Floss[1] and Hans-Josef Kloubert[2]

[1] Technical University of Munich, Germany
[2] BOMAG, Germany

Abstract: During the last few years a very fast development could be noticed in the area of soil and asphalt compaction technology. This is decisively characterized by increasing construction performance, the higher emphasis on economical aspects and the higher demands for quality assurance in earthworks and traffic area construction.
This paper reports about newest innovations in soil and asphalt compaction technology which includes VARIOMATIC tandem rollers and VARIOCONTROL single drum rollers with adjustable vibration, improved heavy padfoot rollers and further development on roller integrated compaction control and documentation systems.

1   Introduction

For the production of stable and deformation resistant traffic areas, dam constructions, waste disposal installations, foundation areas etc. with durable load bearing capacity, machine compaction technology with vibratory compaction equipment is indispensable.
In recent years there have been tremendous developments in compaction technology, mainly influenced by both the further development of measuring and controlling technology and electronic data processing, as well as the changing demands with respect to the application of machines. Construction tasks in modern earth work and traffic area construction are therefore characterised by a higher construction performance, greater significance concerning economy and higher quality demands. Today compaction equipment must be powerful, economical and versatile in use, as well as informing the roller driver about the efficiency of the machine in a simple and understandable way and contribute to quality assurance.

BOMAG recognised demands and development possibilities at a very early stage and already in the eighties thoroughly investigated the compaction processes in the soil during application of vibratory rollers and the interaction between drum oscillation and the soil reaction force changing with increasing compaction (Fig 1).

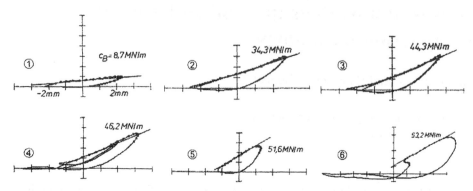

Fig 1: Indicator diagram (soil reaction force versus drum oscillation) with increasing compaction,
BW 213 DH-3

The push in innovation resulting from this led to the development of the Terrameter and the BCM-system. In earth and road construction these roller integrated measuring and documentation systems contribute to compaction control and optimised use of compaction equipment, and at the same time provide a surface covering documentation of the compaction.

In recent years further research and development activities led to the successful market launch of intelligent compaction systems such as VARIOMATIC and VARIOCONTROL, which automatically adapt the most important machine parameter for compaction, the amplitude, to the dominant operating conditions. The controlled roller particularly stands out with respect to compaction performance, depth effect, uniform compaction and its suitability for universal use. Last but not least, the demand for more powerful padfoot rollers for fine grain and rockfill materials, as used on the large scale civil engineering projects of the German Railway, led to the development of new padfoot shapes.

2   Intelligent controllable compaction systems for vibratory rollers

Conventional vibratory rollers are characterised by exciter systems on which the amplitude, the most important parameter for compaction, can only be adapted to the different operating conditions, such as various materials, different lift heights or the changing compaction conditions, by a fixed specification.

With the market launch of the VARIOMATIC system for tandem rollers in 1996 and the VARIOCONTROL system for single drum rollers in 1998 BOMAG was able to perform a remarkable technological leap. Vibratory compaction machines with self controlling compaction system to improve the quality and reproducibility of compaction independently from the operator and to increase the economy of the compaction machine, were available for the first time. This was preceded by several years of intensive research and development activities in co-operation with Prof. Dr. Ing. Floss from the Technical University of Munich and extensive compaction investigations on the test site at BOMAG.

First of all tandem rollers of the 7-8 t class were fitted with the VARIOMATIC compaction system, which has in the meantime been extended to the 10-12 t class. In the range of single drum rollers machines of the 8, 14 and 25 t class are available with the respective VARIOCONTROL system.

The new intelligent self controlling systems VARIOMATIC and VARIOCONTROL are based on the analysis of the interaction between drum and stiffness of the material to be compacted. The compaction energy is automatically optimised by utilising acceleration signals. This adaptation of compaction has the effect that the maximum possible compaction energy is transferred to the construction material at any time, without the disadvantage of the drum changing to jump operation or a destruction of material caused by over-compaction.

3 VARIOMATIC – Compaction System for Tandem Rollers

3.1 System design and working principle

In contrast to conventional systems the exciter of the VARIOMATIC consists of two counter-rotating eccentric shafts generating a directed oscillation (Fig 2). The direction of the resulting force can be changed by offsetting one shaft in relation to the other. Drive and adjustment are accomplished hydraulically. Both shafts are synchronised by a pair of gears. The control of the force direction is accommodated by a control circuit. The system utilises the interaction between the drum and the stiffness of the soil to be compacted, which changes with progressing compaction.

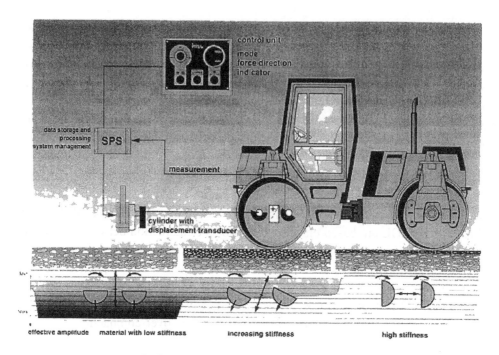

Fig 2: VARIOMATIC system for tandem rollers

Two acceleration transducers on the drum pick up the measuring values and transmit the signals to a processing unit (PLC)[1]. The periodical signals are sensed with a high frequency and permanently compared with a limit value. When the limit value is reached, the PLC transmits a control signal to the control cylinder, which in turn alters the working direction and thereby the effective amplitude (vertical amplitude component), until a perfect adaptation to the respective compaction condition is reached. The system is able to respond very quickly to changing compaction conditions.

Besides the automatic mode described above, in which the system controls itself, it is also possible to pre-select a certain oscillating direction. In this case the roller operator can select from a choice of 6 vibrating directions between horizontal and vertical (Fig 3).

---

[1] programmable logic control

Fig 3: VARIOMATIC control unit for the roller operator to specify the mode of operation

3.2 Applicational advantages of the VARIOMATIC system

The VARIOMATIC system helps the user to improve the performance and quality of compaction considerably and relieves the roller operator from critical application related decisions, so that even less trained drivers can achieve good and uniform compaction results (Table 1).

Since the direction of the directed oscillation is adapted to the travel direction, the result is an excellent flatness of the surface and a lower tendency to scuffing and formation of cracks, because the asphalt is not pushed forward, but is always pulled under the drum. Further advantages are good compaction results, without the risk of grain damage on thin asphalt layers during horizontal vibration, which is similar to oscillation. Horizontal vibration can also be used for the compaction of thin layers on a soft sub-base without loosening and without subsequent compaction of the layer underneath (of advantage when compacting hot on hot).

Table 1: Advantages of the VARIOMATIC concept for tandem rollers

- Universal use on road base and wearing course layers, thin layers and in areas sensitive to vibration effects
- Higher compaction performance without the risk of grain destruction
- Uniform compaction, even on sub-bases with inhomogeneous stiffness
- Better flatness and more uniform surface structure on asphalt layers
- Low tendency to scuffing
- Problem free operation of the roller when rolling joints hot against cold

The VARIOMATIC system offers a particular advantage when being used in urban areas and on bridges, wherever extreme vibration effects for the environment can be expected. Measurements show that the vibration effect on buildings can be considerably reduced when using the possibility to pre-select the vibration (Fig 4).

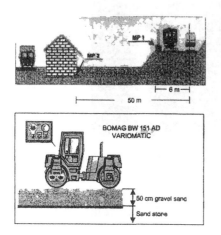

BW 151 AD	Vibration angle	Velocity			
		Measuring point 1 V_{max}		Measuring point 2 V_{max}	
		dB	mm/s	dB	mm/s
Variomatic	90° (vertical)	99	0 89	76	0 06
	72°	98	0 79	76	0.06
	54°	98	0 79	76	0.06
	36°	96	0 63	75	0.06
	18°	96	0 63	73	0.04
	0° (horizontal)	84	0 16	62	0.01
Rail traffic	Goods train	107	2 24	72	0.04
	Express train	100	1 00	69	0.03
	Regional train	94	0 5	68	0.03
Road traffic	Truck			72	0 04
	Comb harvester			69	0 03

V_{max} max. velocity according to Swiss specs: 3mm/sec

Fig 4: Propagation of vibration when working with a 7.5 t tandem roller with VARIOMATIC system

4   VARIOCONTROL compaction system for single drum rollers

4.1   System design and working principle

For the development of the VARIOCONTROL system experiences gained with the VARIOMATIC system in the field of asphalt compaction were used. However, the higher amplitudes needed for earth construction rollers cannot be realised with the exciter system on VARIOMATIC machines. BOMAG therefore developed a new exciter system with directed oscillation, which is able to achieve amplitudes of 2.5 mm and centrifugal forces of up to 500 kN with a vibrating mass of 9000 kg.

Two concentrically arranged vibrator shafts carry three eccentric weights, the two smaller weights near the ends and the larger eccentric weight in the middle of the exciter shaft. The middle eccentric weight rotates in the opposite direction to the outer weights (Fig 5). The resulting centrifugal forces add up to a directed oscillation. The effective direction of this directed oscillation can be adjusted by turning the complete vibrator unit. Any desired angle position between horizontal and vertical oscillating direction is possible. The adjustment of the vibrator unit is accommodated by a hydraulic swashing motor with integrated path measuring system. Similar to the VARIOMATIC system, the VARIOCONTROL system is also equipped with acceleration transducers on the drum, which continually measure the dynamic behaviour of the drum and transmit the data to the programmable logic control (PLC). As an alternative to the fully automated system there is also the possibility of a manual selection from a choice of six vibration directions between vertical and horizontal.

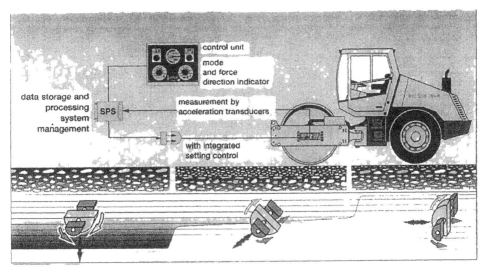

Fig 5: VARIOCONTROL system for single drum rollers

4.2   Applicational advantages of the VARIOCONTROL system

The applicational advantages of the VARIOCONTROL system with directed oscillation concept in comparison with conventional vibratory rollers become apparent particularly in the high compaction performance and depth effect as well as in the excellent adaptability. Fig. 6 shows the generally different course of density over the measuring depth for a directed vibrator and a conventional rotation exciter achieved with the 13 t single drum roller BW 213 D on silty gravel. The directed oscillator already shows during the first passes a better depth effect and a considerably higher increase in density than the rotation exciter with same frequency and linear load. Further passes confirm this trend. The directed oscillator apparently transfers a considerably higher compaction energy to deeper zones than the rotation exciter, the energy transfer of which is more concentrated near the surface of the soil layer, thereby resulting in the distinct formation of a lid.

Light weight penetrometer

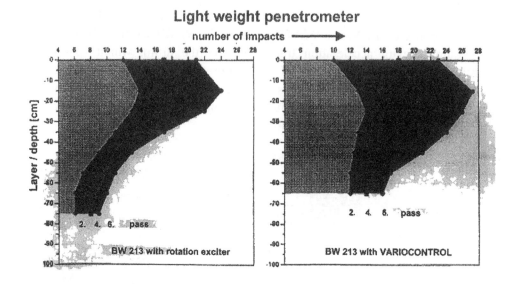

Fig 6: Comparison of the compaction effect of a 13 t single drum roller with conventional rotation exciter and the VARIOCONTROL system with directed oscillator on silty gravel

Further advantages of the VARIOCONTROL system are the adaptability of the compaction process to different stiffness conditions of the layer to be compacted and the respective sub-base conditions. The system enhances the uniformity of compaction, which is of increasing importance in modern road construction and in the construction of foundation areas.

During the construction of frost protection layers and non-bonded wear courses over-compaction with the risk of loosening or even crushing of aggregates is avoided. As with the VARIOMATIC system on tandem rollers the vibration effect on buildings can be minimised by selecting a horizontal vibration direction when using single drum rollers with VARIOCONTROL in urban areas.

5   Special Padfoot Rollers for Cohesive Soils and Rockfill

5.1 Conventional padfoot rollers

For years single drum rollers with padfoot drums have mostly been used on cohesive soils, mixed soils and certain rockfills. These are normally characterised by trapezoidal teeth of 100 mm height with flat side faces and scrapers to keep the drums clean.

In connection with the vibration these padfoot elements produce a kneading and manipulation of materials to reduce air voids and to crush lumps and soft pieces of rock. The padfeet are designed so that the surface area increases in proportion to the penetration into the material thereby effectively adjusting the compaction effect to the stiffness of the soil. Single drum rollers with padfoot drums are available in the weight class from 6 to 25 t and have proven themselves in terms of utilisation and excellent compaction performance (Table 2).

Table 2: Single drum padfoot rollers with performance data acc. to information by the manufacturer

Type of roller and operating weight CECE (incl. ROPS and cab)		t	Compacted layer thickness (m)			
			Rock	Gravel, sand	Granular	Silt, clay
	BW 156 PDH-3	6.5	-	0.40	0.30	• 0.20
	BW 177 PDH-3	7.7	-	0.45	0.35	• 0.20
	BW 178 PDH-3	8.3	-	0.45	0.35	• 0.20
	BW 179 PDH-3	9.0	0.80	0.50	0.40	• 0.25
	BW 212 PD-3	12.0	0.80	0.50	0.40	• 0.25
	BW 213 PDH-3	13.1	0.90	0.60	0.45	• 0.30
	BW 214 PD-3	14.2	1.00	0.70	0.50	• 0.30
	BW 214 PDH-3	14.2	1.00	0.70	0.50	• 0.30
	BW 216 PD-3	16.6	1.20	0.80	0.60	• 0.35
	BW 216 PDH-3	16.7	1.20	0.80	0.60	• 0.35
	BW 219 PDH-3	19.6	1.60	1.20	0.80	• 0.40
	BW 225 PD-3	24.8	2.00	1.50	1.00	• 0.50

PD = padfoot drum   PDH = excellent climbing ability   Roller is particularly suited for this application

Stimulated by compaction technological and performance related problem definitions of earth construction contractors with the compaction of cohesive soils and rockfill on large civil engineering projects in highway, rail track and airport construction BOMAG has lately performed extensive investigations with single drum padfoot rollers. In this context the primary goal was the higher compaction performance on cohesive soil and rockfill as well as the better manipulation and crushing effect on rockfill material demanded by the civil engineering contractors. This experience led to the development of three special padfoot drums, which have already been successfully used (Fig 7).

BW 225 with 150 standard teeth. H=100 mm

BW 225 with 100 pyramid teeth,
H=150 mm

BW 225 with 120 triangular teeth and additional
hard-faced tips, H=200 mm

BW 225 with 100 triangular teeth and cutting
plates in between, H=200 mm

Fig 7: 25 t single drum vibratory rollers with various types of padfoot teeth

5.2 Rollers with pyramid teeth

Compared with conventional padfoot drums, drums with pyramid teeth are characterised by higher teeth, considerably smaller tip areas and steeper tooth flanks (Fig 8). Due to the higher specific pressure pyramid teeth penetrate deeper into the soil and achieve a more intensive kneading and compaction of the soil. An additional benefit is the better self-cleaning effect between the teeth on cohesive soils. Clogging of the drum, which is disadvantageous for the compaction effect, is avoided. During the civil engineering work at the new airport construction in Leipzig rollers with pyramid padfoot teeth were able to exceed the compaction performance of conventional padfoot rollers on boulder clay considerably. With identical static weight and number of passes the special pad foot roller was able to compact an on average 10 cm thicker layer, because of its deeper penetration.

Fig 8: Higher compaction performance on boulder clay by using a 25 t roller with pyramid padfoot teeth. Leipzig Airport

5.3 Rollers with triangular teeth

For the subsequent crushing of very hard and brittle pieces of rock, as they are found in the quarry or excavation material used on the new ICE rail track construction Cologne-Rhine/Main for the construction of dams, BOMAG equipped a 25 t single drum vibratory roller with triangular teeth (Fig 9). When using single drum rollers with vibration on coarse, solid rockfill the tips of the triangles produce such high pressure and splitting forces, that the desired crushing and edge fracturing of the coarse grain and therefore an intensive compaction of the packing bed is achieved with a considerable reduction of air voids. In order to withstand the excessive loads acting on the triangular teeth, their tips are made of a wear resistant material.

Fig 9: Higher crushing effect of a vibratory roller on quartziferous sandstone achieved by a special padfoot drum (triangular teeth), new ICE rail track construction Cologne-Rhine/Main

5.4 Rollers with triangular teeth and cutters in between

25 t rollers with triangular teeth and cutters in between are most suitable for the crushing of rockfill material (Fig 10). The coarse grain material is split by the triangular teeth and subsequently crushed by the cutters. At the same time the cutters prevent jamming of crushed material between the teeth.

In the area of construction lot A of the ICE rail track Cologne-Rhine/Main quarry and excavation materials of various compositions and with different degrees of weathering and hardness are used for the construction of dams with heights of up to 18 m. This includes also grain sizes of more than 200 mm, which are not permitted for these dams. With the use of special padfoot rollers the highly expensive crushing, screening and manual sorting of coarse pieces is no longer necessary. These special padfoot rollers are able to crush and effectively compact laminated fragments of clay or silt slates and hard quartzious sandstone of up to 500 mm in size.

Fig 10: Compaction of dam construction material in 30 cm layers and crushing of oversize
components with special padfoot drum (triangular teeth with cutters in between)

6   Compaction Measuring and Documentation System BTM/BCM

6.1   Status of technology

The basis of the compaction measuring system Terrameter BTM developed by
BOMAG is the interaction between the acceleration of the roller drum and the
stiffness of the soil changing with progressing compaction. In conjunction with the
BCM documentation system this enables a direct conclusion from the dynamic
behaviour of the roller drum on the compaction status of the soil (Fig 11). This
results in considerable quality assuring and economical benefits for the
construction operation. The roller passes are thereby reduced to a minimum. Soft
spots can be identified and treated immediately. This causes as little interference
for the construction operation as possible and the uniformity of compaction is
enhanced in a way which would not possible when using conventional, punctually
applied testing methods.

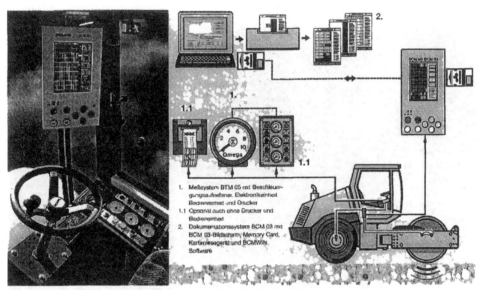

Fig 11: BOMAG compaction and documentation systems

Terrameter BTM

The drum is fitted with a detector unit with two acceleration transducer arranged vertically to each other, which pick up the acceleration of the drum. In the electronic unit a micro processor stores the data transmitted by the detector unit. An analogue display, two control lights and a printer are provided for the output of measuring values. The roller operator can read the actual, non-dimensional measuring value with the designation "Omega-value" on the analogue display, while the compaction is in progress. Moreover, the Omega-values for a certain measuring distance, which may range from 0 to 1000, depending on status of compaction or stiffness, can be viewed after each pass in form of a line or bar chart. Two control lights on the control unit informing the roller operator whether further effective compaction passes are possible or not, are particularly helpful. For this purpose the measuring system compares the mean Omega value of a forward (reverse) pass with the value of the previous forward (reverse) pass. When falling below a determined difference in values the red control light will light up, indicating the end of effective compaction work.

BOMAG Compaction Management BCM

During the compaction process the measuring values collected by the BTM are graphically and numerically displayed for the operator on the colour screen of the BCM documentation system and analysed, managed and documented on a PC by the evaluation program BCMWIN. The data transfer between display and PC is accomplished with the help of a Memory Card. This serves also for the communication between construction supervision and roller operator. The area to be processed is divided into a raster of rolling tracks and raster size, positioning in the field by xy-co-ordinates, calibration results and other data for the description of the compaction work are prepared on a PC with the BCMWIN program and written to the Memory Card.

6.2   Using the measuring and documentation system

Particularly in large-scale construction projects, such as airport, rail track and highway construction, the BTM and BCM systems have been used most successfully. The experience made in both the assessment of the sub-base with respect to uniformity of its load bearing ability and the compaction of frost protection and non-bonded wearing courses were to a wide extent positive with respect to economy and quality assurance.

Since the principle of function is based on the interaction between the acceleration behaviour of the vibrating drum and the dynamic stiffness of the soil changing with progressing compaction, the measuring values are influenced by both the soil specific as well as the machine specific parameters.

The soil specific influential parameters include those that describe the stiffness, i.e. the strength and deformation characteristics of a soil. These are mainly grain size distribution, plasticity, water content and density.

Non-cohesive soils
For non-cohesive soils there is a direct relationship between density and stiffness. With increasing density the Omega values also rise because of the increasing stiffness. Since there is a distinct assignment between measuring value and modulus of deformation and therefore also between measuring value and density or degree of compaction, it is possible to calibrate the Omega value to the modulus of deformation or the degree of compaction (Fig 12).

Cohesive mixed soils
On cohesive mixed soils the influence on the stiffness of the soil increases with a rising water content. For water contents below the optimal Proctor water content this influence is still relatively small, so that in such a case a direct relationship between Omega-value and density or degree of compaction is also possible.

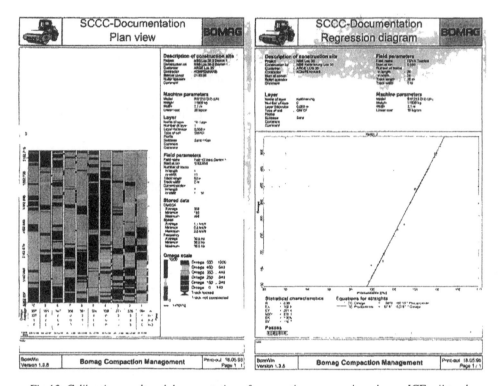

Fig 12: Calibration result and documentation of compaction on gravel sand, new ICE rail track
construction Cologne-Rhine/Main

The influence on the stiffness increases with a rising water content. A calibration is
more difficult, because measuring value and density or modulus of deformation
can no longer be assigned without considering the water content.

However, in these cases the measuring values still permit a qualitative assessment.
Inhomogeneities and weak spots caused by extremely different water contents can
be localised and verified (Fig 13).

Cohesive soils

On cohesive soils a direct calibration is normally no longer possible, because the
deformation characteristics of these soils depend very strongly on the water
content.

However, on cohesive soils with constant water content and even composition, as
used as sealing material on sanitary landfill sites, the relationship between Omega-
value and density or modulus of deformation can in individual cases be used as a
criterion for quality.

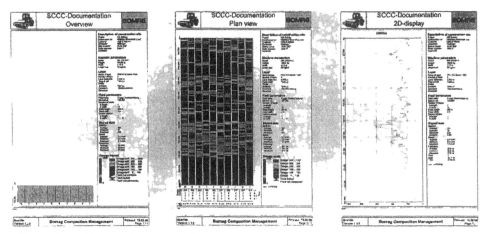

Fig 13: BCM 03 measuring protocol for the verification of the uniformity of compaction on mixed soils. new ICE rail track construction Cologne-Rhine/Main

6.3   Outlook on further developments

The further development of roller integrated measuring and documentation systems includes the location of the roller by means of satellite navigation (GPS), the cross-linking of position data with compaction data and the real-time presentation of the compaction results in a 3D-model. Apart from this reading interfaces for DGM planing software (Digital Area Models) are planned, so that actual positions can be compared with the specified data of the DGM during the compaction process. A first demonstration of the GPS possibilities was performed by BOMAG during the compaction equipment exhibition BAUMA 98 in Munich (Fig 14).

Another very promising perspective of the surface covering dynamic compaction control is the development of the VARIOCONTROL system for single drum rollers. All previously used roller integrated measuring and documentation systems show non-dimensional characteristic values related to the stiffness of the soil, as long as they were generated with constant machine related parameters. This means that amplitude, frequency and travel speed must not change during a measuring pass. During the development of the VARIOCONTROL system a method for a reliable direct detection of the dynamic stiffness of the soil was found, which works independently from machine related parameters. Fig. 15 shows the relationship between the dynamic modulus of deformation E_{dyn} and the position of the directed vibrator of the VARIOCONTROL system.

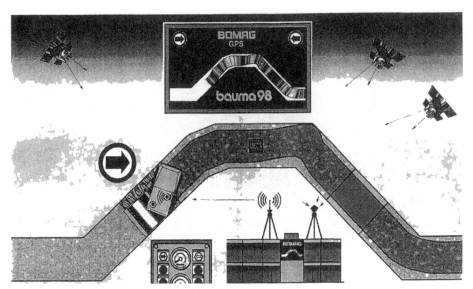

Fig 14: Location of the roller by GPS and real-time presentation of the compaction results

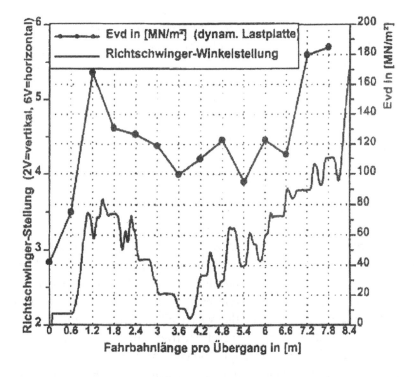

Fig 15: Relationship between the dynamic modulus of deformation E_{dyn} and the position of the directed vibrator

The function principle of the VARIOCONTROL system makes it possible to determine the stiffness of the soil by the resulting position of the directed vibrator. With increasing stiffness of the soil the oscillating direction is adjusted from vertical towards horizontal.

References

[1] FLOSS, R. (1992): *Flachendeckende Qualitatskontrolle bei Verdichtungsarbeiten im Erd- und Straßenbau.* Proc. 1. Int. Symposium „Technik und Technologie des Straßenbaus", BAUMA 1992

[2] FLOSS, R. (1996): *Qualitätssicherung im Erdbau - Anwendung der neuen Prufmethoden gemaß ZTVE-StB 94.* Forschungsgesellschaft für Straßen- und Verkehrswesen, Köln, Schriftenreihe der Arbeitsgruppe „Erd- und Grundbau", Heft 7

[3] FLOSS, R., HENNING, J. (1998): *VARIOMATIC. Ein entscheidender Schritt zur Qualitatssicherung im modernen Erd- und Verkehrswegebau* BOMAG, Heft BA 049, Boppard

[4] KLOUBERT, H.-J. (1993): *Flachendeckende dynamische Verdichtungskontrolle als Beitrag zur Qualitatssicherung im Erd- und Straßenbau* BOMAG, Heft VPA 3/93, Boppard

[5] KLOUBERT, H.-J (1999): *Anwendungsorientierte Forschung und Entwicklung lost Verdichtungsprobleme im Erd- und Straßenbau* 28. VDBUM Seminar, Seminarband 1999, Verband der Baumaschinen-Ingenieure und Meister e.V., Stuhr

[6] KROEBER, W. (1999): *VARIOCONTROL und FDVK im Erdbau - schwierige Verdichtungsaufgaben sicher und wirtschaftlich gelost.* 28. VDBUM Seminar, Seminarband 1999, Verband der Baumaschinen-Ingenieure und Meister e.V., Stuhr

Soil improvement by vibro compaction in sandy gravel for ground water reduction

N. Sidak

Keller Grundbau Ges.m.b.H.. Wien

When about 65 years ago. the company Johann Keller received the German patent for improvement of granular soils by means of depth vibrators applying the vibro compaction method invented by Keller, nobody could foresee the developments soil improvement would take within the coming decades.

My lecture today refers to soil improvement works in non-cohesive soils executed by means of vibro compaction. In this context it has to be remarked that the further development - also the soil improvement of cohesive soils - the so-called Keller Vibro Replacement, is of great significance today.

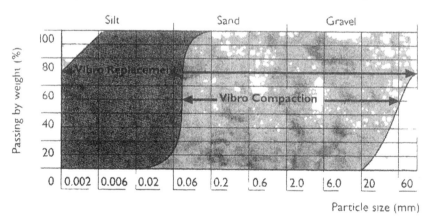

Fig. 1 Table of Application Scheme

In the course of discussions regarding a new flood protection scheme for the Danube river in Vienna, observations were made by the Republic of Austria - represented by the Federal Office for Water in Vienna and the special departments of the Municipal Building Control Office of the province Vienna - to rehabilitate the existing flood protection dams along the river Danube east of Vienna. These dams had been built in the eighties of the preceding century. These works were supported and supervised at the time by the Technical College of Vienna for various research projects.

In 1968 Keller together with the above-mentioned institutions developed a process for the reduction in permeability of sandy gravels in areas of existing flood protection dams.

Flood embankment on the left danube bank

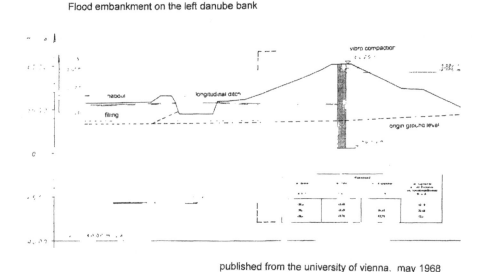

published from the university of vienna. may 1968
geotechnical engineering institut

Fig. 2   Cross-section of a flood protection dam

Aim of these works was to stabilize the dam by reducing the voids and by compacting as well as flushing the existing dam with sand in order to change the permeability and the saturation line in such a way, that in case of high water the security of the dam would not be at risk. The deep compaction process was started at the existing dam crest in intervals of approx. 1 m. These trials showed very good results with the consequence that the Municipality of Vienna and also the Federal Water Authority as representative of the Republic decided in the course of the coming years to execute extensive works to reduce the permeability of sandy gravels at existing flood protection dams. These works were carried out primarily at the Danube river.

The company named Johann Keller at that time, contributed significantly to the development of this very economical as well as environmentally friendly procedure. This led to the further development of flood protection systems, which also have to serve the purpose of protection of human beings.

For the execution of the dam protection works at the river Danube in the 1970's it was necessary to rebuild the whole sewerage system along the river Danube by constructing large collection sewers to the main water treatment facility which was to be built in the Simmering area of Vienna.

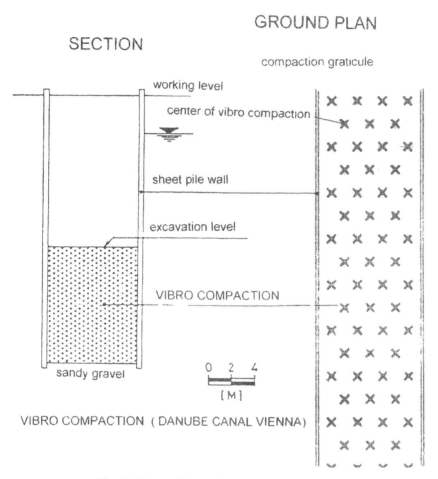

Fig. 3 Scheme of Vibro Compaction at the canal in Vienna

Formerly bottom seal slabs in ground water were usually executed by grout injections. In order to achieve cost savings Keller - at that time GKN Keller - developed an alternative method using deep compaction. In this method, the sandy gravel found below the bottom of the excavation is densified by compaction and introduction of sand in such a manner that its permeability is significantly reduced. Following this, various structures in the sewerage system and canals could be built using open dewatering.

For the planner the problem arises that due to the presence of very pervious sandy gravel and a very deep backwater level, an economically effective sealing cannot be achieved.

The depth of the retaining wall, generally a sheet pile wall, but which could also be a diaphragm wall or a Soilcrete wall - will result from the static requirements for the stability of the sheet pile wall.

Planning as well as execution show that the static requirements can be solved without any problems by using wall bracings in the zone of the sewer cross-section.

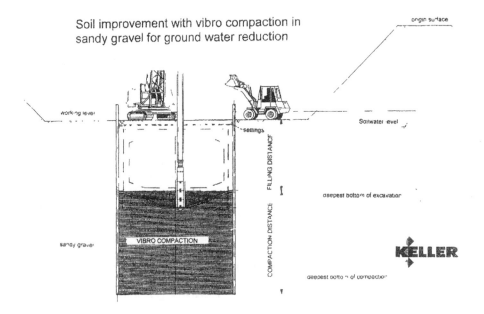

Fig. 4   Deep Compaction to reduce permeability

The execution (see Fig. 3) takes place from the pre-excavation level which corresponds to the driving level of the sheet pile wall and is approx. 1.0 m above ground water level. This is the level from which the deep compaction will take place, the depth of which will result from the required level difference or the bottom of the sheet pile wall. This means the deep compaction process takes place within an impervious construction pit girding between the bottom of the compaction and the deepest excavation slab. The zone between the deepest excavation base and the working level was treated as the so-called "intermediate zone" and usually only loosely compacted. Then excavation of the pit can be started (see Fig. 4).

Certainly a technically sound and economically justifiable solution using the deep compaction process with the aim to reduce the permeability depends on the required height of the ground water lowering. In most cases this method is economical and technically feasible in case of differences in ground water levels of up to 4 m.

In order to limit the area of each section, cross sheet pile walls with appropriate lock jaws are to be foreseen so that each ground water lowering section does not have an area larger than 1000 m².

For the execution of the works, thorough soil investigations have to be made which give information about grain size distribution and the prevailing permeability coefficients of the in-situ sandy gravels.

Requirement is that the method is only a temporary measure and after completion of the works (removal of the excavation supports and stoppage of ground water lowering), the original ground water levels and flow are restored. The only change is that the vibro compacted zone made up of sandy gravels now has a lower permeability.

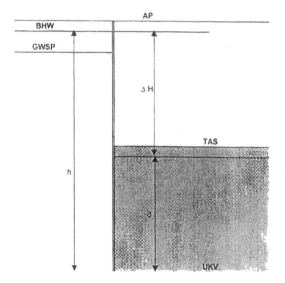

AP - working level
TAS - deepest excavation level
BHW - highest water level
Δ H - lowered water level
UKV - deepest compaction level
GWSP - ground water level

The design must fulfil two requirements: firstly the required safety for hydrostatic uplift of the excavation base and the safety against base failure of the excavation pit, secondly the requirement for the decrease in the permeability or the allowable lowering of the ground water level outside the construction pit. The depth and height of compaction and the required reduction in permeability are to be designed to meet the above requirements. In addition, when determining the depth the flood level has to be taken into consideration.

When preparing the design, it has to be ascertained that for a certain high water level of a structure the condition of the structure has to be taken into account. From this calculation the quantity of residual water after pumping out the excavation pit is revealed, taking the size of the area to be treated into account. After having reached the high water level of the structure the safety of the uplift is no longer guaranteed and as a consequence in such a case the construction pit has to be flooded.

During the planning stage the criteria for the maximum ground water lowering have to be determined exactly. This also applies in case of the normal bottom seal slab, which is designed for a certain water pressure.

Instead of using grout material sand or fine sand is filled and compacted in the Keller vibro compaction method to reduce the permeability of sandy gravels.

Assuming a starting K-value of 10-3 m/sec, the permeability is reduced by up to two decimal powers, which proves to be an economical solution for construction sites to be kept free of water for short periods. In addition, the improvement attained with regard to soil parameters also allows higher ground pressures and hence a more favourable design of the earth retaining system due to the increase in the passive resistance of the soil.

The entire compacted volume beneath the bottom of the construction pit is used for the reduction in the hydraulic gradient. With regard to weight, the bouyant unit weight of the soil is considered. The increased unit weight following compaction allows a lower embedment depth of the retaining wall.

Penetration of the depth vibrators is usually assisted by water flushing. That means, sufficient water as well as the appropriate pressure pumps have to be made available during execution of the works. After having reached the final compaction depth sand is added. The sand is preferably chosen according to the missing grain sizes in the in situ soil.

Grain Size Distribution Lines

GRADING CURVE

Probe 1: natural (origin) gravel

Probe 2: added material

Probe 3: through vibrocompaction mixed soil

Probe 4: soil mixed in the laboratory

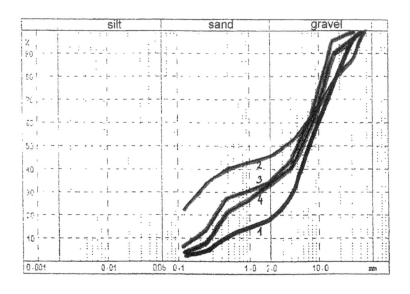

The fill material is transported to the vibrator by means of a wheel loader and is then added in portions via the sand chutes close to the vibrator.

During the deep compaction process approx. 15% of the soil volume can be vibrated in and installed as fill material. By the addition of appropriate sand material using a depth vibrator, the grain size distribution of the in situ is changed in such a way that the permeability of the prevailing soil is considerably reduced.

Due to the hydraulic gradient, the filter resistance of the mixture as well as a resistance against inner erosion (suffusion) is necessary. An analysis for the verification of the safety against base failure is to be carried out.
The residual water volume after dewatering the construction pit is estimated based on the above analysis.

After having excavated the pit completely and after having emptied the pit by means of pumps, the water quantity streaming from the subsoil, including leakage water, is pumped using a dewatering system comprising of vertical and horizontal drain pipes via appropriate pump wells. This measurement requires routine and appropriate supervision, as is the case with each open dewatering.

Applying the Keller deep compaction method, the quantity of water to be pumped can be reduced to a tenth or even to almost a third, depending on the water pressure. Also the seepage rate in the outflow zone is reduced to approx. the power to ten, which leads to an essential improvement of the hydraulic stability. The danger of inner erosion is drastically reduced by the application of deep compaction, which results in an improvement of the following soil mechanical properties:

- homogenisation of the heterogeneous in-situ material
- flushing of sand into the grain skeleton of the in-situ gravel
- an increase in density
- an increase in the shear parameter and
- equalising the grain distribution line.

It is provided, however, that an appropriate outfall is available for pumping the water.

After having installed an appropriate filter layer of filter concrete or filter gravel, the base foundation course is added and then the reinforcement and concreting of the slab and the walls is carried out.

In this way, ground water lowering in long excavation pits is carried out by dividing it into sections and then restoring the ground water to its natural level following completion of works.

This development has lead to the fact that Keller has applied this method successfully to numerous structures, which have been founded in ground water, e.g. buildings, sewage treatment works with large basins and road construction projects like underpasses. This method has been used as an alternative to injection methods.

The spacing between the compaction points (see Fig. 3) is in this case 2 m in the sandy gravels.

The spacing cannot not be increased or decreased arbitrarily. The grid spacing is not determined by the strength of the depth vibrator, but by the amount of sand which has to be introduced and compacted. The aim is to construct a homogeneous compacted mass and most important of all is the introduction of the sand into the sandy gravel, which has become weightless during the compaction process.

Finally some comments on verification and control of the works carried out:
The final verification can only be done by measuring the actual quantity of residual water to be pumped out. First this is done by making an estimation based on an arithmetical calculation. Furthermore, it is attempted to furnish proof beforehand by using other forms of calculation.

One of these is the determination of the permeability of the soil mix based on the permeability of the in-situ sandy gravel and the permeability of the added sand. These mixtures can easily be produced in a laboratory or samples can be taken from the excavation pit on site and then investigated. Based on the permeability coefficients determined in the laboratory, the residual water quantity to be expected can be estimated quite precisely.

The third solution is to use in-situ tests executed according to ÖNORM B4422, field method, which in recent times have been applied with great success in the area around Vienna as well as in Tyrol. When using this method, seepage tests in compacted and non-compacted soils are being carried out which allow the determination of the relative improvement before hand.

In the Tyrol area (Inntal) Keller has executed various projects successfully in the past few years, i.e. sewage treatment works, underpasses of highways and flood protection schemes in ground water, applying the afore-mentioned methods.

References

Borowicka, H. (1968): Der Hochwasserdamm am linken Donauufer im Wiener Bereich.

Stockhammer, P. / Baumann, V. (1976): Erfahrungen mit Abdichtungssohlen bei Baugruben im Grundwasser.

Kirsch, K. (1979): Erfahrungen mit der Baugrundverbesserung durch Tiefenruttler.

Bartels, K. / Jebe, W. (1983): Entwicklung der Verdichtungsverfahren mit Tiefenruttlern von 1976 – 1982.

György, L. (1986): Anwendung von Tiefenrüttlern und des Soilcrete-Verfahrens für die Verbesserung der Standsicherheit des Donaudammes in Albern bei Wien.

Sidak, N. (1993): Sohldichtung mittels Tiefenverdichtung

The in-situ densification of granular infill within two cofferdams for seismic resistance

B.C Slocombe [1], A.L Bell [1], R.E May[2].

1   Keller Ground Engineering. UK.
2   Gibb Ltd, UK.

Abstract:

The paper discusses the construction of two 18.4m diameter cofferdams in 15m of water. The cofferdams were designed for a 1 in 10,000 year earthquake with a peak horizontal ground acceleration of 0.25g. The cofferdam fill was a nominal 6mm to 3mm aggregate which, being hydraulically placed through water, resulted in an initial low relative density. The fill has several performance requirements including adequate weight to generate the necessary sliding resistance on the base of the cell and, in the seismic case, adequate density to avoid liquefaction.

The minimum required relative density for this design was 80% using the Baldi approach. In-situ densification was achieved using the Vibro Compaction technique with large 125kW vibrators. The sheet piling was instrumented to check lateral pressures and avoid over-stressing of the sheet pile interlocks. The instrumentation provided valuable insights into the response of the cells to the sequence of treatment. Confirmation that the necessary density had been achieved was measured by electric cone test. A particularly interesting aspect of this project was presented by the opportunity to compare the results of two specialist contractors and their equipment and methodology. Post-treatment CPT results clearly illustrate the effects of the different compaction strategies.

Introduction

Two cofferdam cells, designated East and West, were constructed as a temporary measure to protect other works to be performed within a dry dock. Each cell consisted of 140 Frodingham SW1-A grade 50 sheet piles, support by an internal framework, which were lifted into place by floating crane. The 19.0m long piles were then driven through a gravel blanket to the bedrock or an existing concrete surface.

The cells were filled with granular material, compaction performed and construction of the seals between the two cells and adjacent basin walls completed. Strain gauges and inclinometers were used to monitor the piles for tensile hoop strain and deflection respectively throughout filling, compaction and commissioning stages of construction. A weighted blanket was constructed on the down stream side of the cofferdam to provide security against piping.

On completion of the filling and Vibro Compaction, the cells were commissioned by dewatering the area on the dry dock side of the cofferdam.

Selection of materials

The specifications for the cofferdam mattress, laid on the basin floor before placing the cofferdam, and the cell fill were similar, the materials being required to be well graded and angular to sub-angular. The initial specification is presented in Table 1:

Table 1: Specification Parameters

Bulk density (Mg/m³)	2.00
Dry density (Mg/m³)	1.80
Shear strength (c^1/ϕ^1)	0/37.5°
Maximum particle size (mm)	60
Minimum particle size (mm)	<5% passing 63μm
Maximum passing 2mm	20%
Permeability (m/sec)	>10⁻³
10% Fines Value (kN)	40

Testing was required in accordance with the relevant BS 1377 (1990) and BS 812 (1990) methods with five tests per cofferdam plus three on the mattress fill for maximum and minimum densities and particle size distributions, with a minimum of one test per 300 tonnes of fill.

Various sources of potential fill were considered and a blended crushed limestone aggregate selected from a local quarry. The fill was blended from the following standard particle size ranges:

    a)   6mm single-size

    b)   3 to 5mm size

    c)   3mm to dust size

Preliminary laboratory tests revealed that whilst a 50/50 combination of types a) and b) had ample strength and permeability properties a high relative density would be required. The addition of 8 to 10% of type c) was therefore required to increase the density. Limits of 30/60/10 to 60/30/10 combinations were found to exhibit acceptable properties. The final mix was a slightly sandy fine gravel. Test results for various gradings are presented in Table 2:

Table 2. Initial Testing of Fill Material Blends.

Fill blend %	46/46/8	30/60/10	60/30/10	50/50/0
Maximum dry density (Mg/m³)	1.87	1.72	1.74	1.80
Minimum dry density (Mg/m³)	1.33	1.41	1.41	1.45
Particle density (Mg/m³)	2.655	--	--	--
Drained strength c^1 (kPa)	68	9	20	17
ϕ^1 (degree)	44	43	42	48
Permeability (m/sec)	4×10^{-2}	3×10^{-3}	4×10^{-3}	9×10^{-3}
At dry density (Mg/m³)	1.47	1.66	1.75	1.58

The design was therefore based on conservative parameters of:

Minimum dry density = 1.41 Mg/m³
Maximum dry density = 1.68 Mg/m³
Specific Gravity   = 2.65 Mg/m³

These values revealed that in order to obtain the required design saturated bulk density of 2.00 Mg/m³, a relative density of 77% was required. A minimum value of 80% was therefore specified, to be proven by Standard Penetration Tests using the Stroud (1989) correlation and/or CPT's using the Baldi (1982) correlation.

Instrumentation

The filling, compaction and commissioning operations were monitored using strain gauges fixed to the sheet piles, inclinometers and piezometers (see Figure 1). The level of fill around the inner circumference of each cell was also monitored during filling and compaction. The instruments were read at intervals of :

Strain gauges   – Every 15 minutes (by datalogger)
Inclinometers   – Before and after each filling and compaction operation
Piezometers   – Before and after each filling and compaction operation
Fill level   – Every 30 minutes

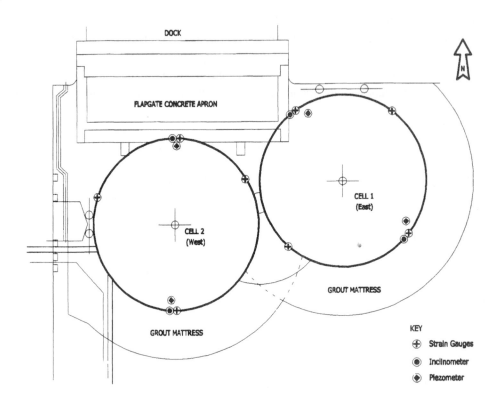

Figure 1 General layout

The strain gauges were fixed in pairs to the outer face of the sheet piling and equipped with a protective cover. At each instrumented location the strain gauges were placed 1.0 m, 2.5 m, 4.5 m and 9.5 m above the toe of the pile, designated locations 4, 3, 2 and 1 respectively.

The principle purposes of the strain gauge data were to check the lateral pressures acting on the inner face of the cells and to monitor the clutch interlock forces. The gauges recorded strains due to hoop tension in the cells and due to bending of the individual sheet piles. The hoop tension was assessed as contributing 44% of the total measured strain for the instrumentation arrangement adopted.

The ultimate clutch interlock capacity for the sheet piles was 384 tonne/linear metre and a minimum factor of safety of 2 was required under permanent loading conditions. The working loading of 192 tonne/linear metre equated to a gauge strain of 1605 microstrain. A limit of 1025 microstrain was set during the compaction operations to allow for the calculated increase in strains during the subsequent commissioning works.

Placement of fill

The sheet piles were required to be driven to a high degree of verticality. This demanded that the bed of the cofferdam mattress was level. The surface of the basin was first removed of silt and loose slate using a Toya suction pump and grab for clean exposed rockhead. A survey revealed the bed level to vary by 2.7m over the cell areas. The cofferdam mattress was then placed using a levelling frame, with screeding beam used to achieve the required level tolerances. This resulted in a mattress thickness of typically 0.9m but up to 2.5m maximum thickness. The sheet pile frame was then lifted into position, individual piles unbolted, and driven through the mattress into the underlying slate bedrock.

Fill placement within the cofferdam was achieved by a pump barge carrying a hopper which mixed the aggregate with water and pumped the mixture via 600mm diameter pipeline into the cofferdam cell. The initial discharge into the West cell was performed with the discharge point close to the west side of the cell. This resulted in uneven filling and the arrangement was modified to provide downward discharge in the centre of the cell. The cells were filled in five or six lifts to about 3.0m above the dock water level.

Further tests on the fill materials recorded:

Minimum dry density = 1.52 to 1.59 (Mg/m³)
Maximum dry density = 1.85 to 1.94 (Mg/m³)
Specific Gravity = 2.68 to 2.72 (Mg/m³)

The strains recorded during the filling of the West Cell are shown on Figures 2. Figure 3 shows the average strain distribution with elevation for the West Cell at completion of filling.

Compaction of fill

Hydraulic placement of fills through standing water frequently results in low relative density. Compaction of the fill by Vibro Compaction (sometimes called Vibroflotation) was therefore proposed, followed by placing and compaction of the uppermost layers by conventional vibratory drum roller.

The initial compaction was performed by the Main Contractor using hired 90 kW hydraulic vibrator and experienced operator. Initial trials were performed on a 3.0m grid and tested by dynamic penetrometer. Two stages of treatment were then

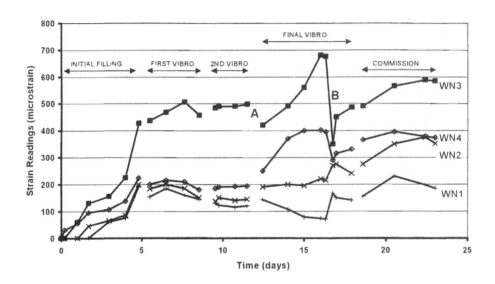

Figure 2. Recorded Strains for the West Cell with Time

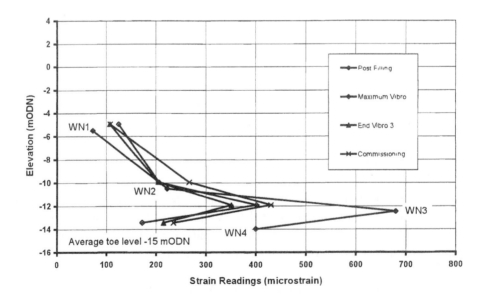

Figure 3: Mean and Peak Strain Distributions for West Cell

performed on a 2.5m offset square grid to depths of up to 16m and tested by boreholes with SPT's at 1.0m intervals, the second stage also with cone penetration tests. These failed to achieve the specified bulk density for the fill, equivalent to a minimum relative density of 80% or $(N_1)_{60} = 38$ (Stroud).

Details of the Stage 1 and 2 average reductions in fill levels and average post-treatment relative densities for the West and East Cells are given in Table 3 and the strains generated are shown in Figure 2. Point A in Figure 2 shows a redistribution of the strain levels between the completion of Stage 2 and the commencement of Stage 3. This was caused by excavation and replacement of the mattress on the north side of the cell.

The compaction results for Stage 2 were particularly poor below about 10m depth, as illustrated in Figure 4, so a third attempt was made. The vibroflot however met with refusal at 12m depth and a specialist Vibro Compaction contractor, Keller Ground Engineering, was approached. After inspection of the site, review of information and interview, they undertook to compact the fill to the required minimum 80% relative density with target average 85%, under subcontract to the Main Contractor.

The stage 3 compaction was performed using the 125 kW Keller S-type depth vibrator, powered by 250 KVA diesel generator. In contrast to hydraulic vibrators, the power is rated within the nosecone and develops high centrifugal forces in a horizontal plane at a frequency of 30 Hz at the main area of contact between fill and vibrator.

The basic unit weighed 2.4 tonnes and, in recognition of the density already attained to the top 10m fill depth, additional heavy-weight extension tubes for 26m total length with high capacity water pumps were provided to ensure that the full fill depth would be re-treated. The much more powerful vibrator was suspended from a very large pontoon crane.

Treatment was performed using a 2.45m triangular grid and, to avoid excessive vibration, minimum distance from sheet piles of 2.0m. The vibrator was also surged at various levels during penetration to create an annulus between fill and vibrator. This enabled cofferdam fill heaped at the surface to penetrate to the base to enhance compaction by both densification and displacement at depth. The vibrator also penetrated the underlying gravel mattress down to bedrock.

As with the previous stages, the uppermost fill was first loosened using a backhoe. Compaction then commenced to the southern section of the East Cell. Treatment

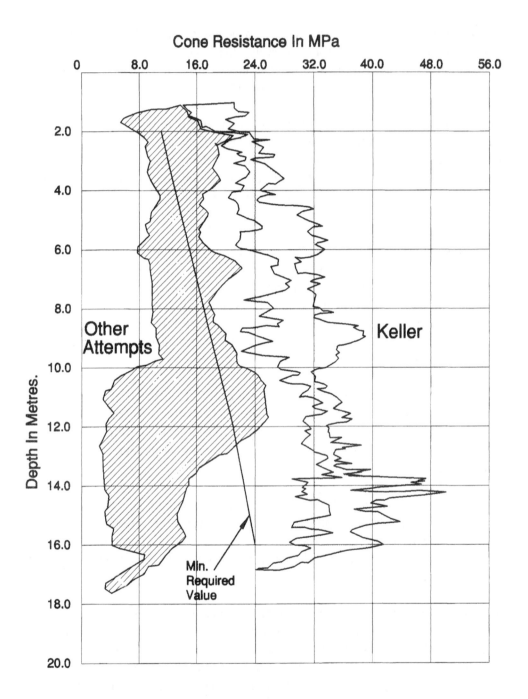

Figure 4: Comparison of post-treatment CPT's of Vibro Compaction
by other attempts and Keller Ground Engineering

was then transferred to the southern side of the West Cell but was stopped two days later when strain readings on gauges on the northern side of the cell reached a value of 675 microstrain (Point B on Figure 2 and Figure 3). After careful analysis of the strain data to date, and associated patterns and trends, the compaction continued adjacent to the northern edge of the cell. This resulted in the recorded reduction in strain levels and the treatment was completed without any further interruption. The coefficient of lateral earth pressure (K) peaked at 0.56 and reduced to 0.40 by the completion of the compaction stage.

Surface levels were measured as having reduced during the Stage 3 compaction by almost as much as to Stages 1 and 2 compaction combined, with over 2.0m being recorded near to the sheet piles. The reductions in stress caused by the reductions in fill levels during compaction had a particular effect on the upper strain gauges (WN 1 and 2) as can be seen in Figure 2.

Post-treatment cone penetration tests were performed at six locations in each cell (see Figure 4). These demonstrated average relative densities in excess of 90% for the West Cell and 87% for the East Cell and hence the specified saturated bulk density requirement had been achieved.

Table 3: Induced reduction in fill levels (m)

Stage	West Cell	Dr %	East Cell	Dr %
1	1.34	69	1.43	63
2	0.36	70	0.74	74
3	1 78	90	1.78	87

On completion of the Vibro Compaction and testing, the fill levels were made up to +4.75 mODN with conventional vibrating roller compaction. The cells were commissioned by pumping out the space between the cofferdams and the dry dock gate while monitoring the various instrumentation. The strain gauges recorded a modest increase on the north side of the cell as shown in Figures 2 and 3.

Conclusions

a)   Seismic design of cellular cofferdams requires high bulk density in the fill to resist potential liquefaction. Advantage can then be taken of the high bulk density to optimise the cofferdam size.

b)   It is essential that potential sources of selected fill be subjected to laboratory testing for suitability in advance and during placement.

c)   Vibro Compaction has been successfully used to achieve relative densities in excess of 80% in a gravelly fill. This project has demonstrated that care is required in the selection of suitable equipment, sequence of operation and choice of experienced contractor.

d)   The use of strain gauge instrumentation has enabled the filling and compaction operations to be controlled, thus avoiding interlock tension failure of the sheet piles.

References

1   Baldi, G, Bellotti, R, Ghionna, V, Jamiolkowski, M. and Pasqualini, E. (1982): *Design Parameters for Sands from CPT*, 2nd European Symposium on Penetration Testing, Amsterdam, 425 to 432.

2   Stroud, M.A. (1989): *The Standard Penetration Test - Its Application and Interpretation*, Penetration Testing in the UK, Birmingham, Thomas Telford, 29 to 49.

3   Kirsch, K. (1985): *Application of Deep Vibratory Compaction in Harbour Construction*, Egyptian-German Seminar in Cairo.

4   Bell, A.L, Slocombe, B.C, Nesbitt, A.M. and Finey, J.T (1986): *Vibro-compaction Densification of a Deep Hydraulic Fill*, Building on Marginal and Derelict Land, Glasgow, Thomas Telford, 791 to 797

5   Castelli, R.J (1990): *Vibratory Deep Compaction of Underwater Fill*, Deep Foundation Improvements: Design, Construction and Testing, Las Vegas, ASTM STP 1089, American Society for Testing and Materials, 279 to 296.

6.   Priebe, H.J. (1995): *The Design of Vibro Replacement*, Ground Engineering, 12, 31 to 37.

Vibro Compaction for the Lock Hohenwarthe

Paul Scheller

Bauer Spezialtiefbau GmbH, Schrobenhausen, Germany

Abstract: The Lock Hohenwarthe near Magdeburg, Germany is the key structure for the intersection of the River Elbe with the Mittelland-Canal and the Elbe-Havel-Canal. Former construction started in 1939, but was terminated in World War II. The unfinished structure has been covered without any compaction using locally available sand. The area of the old ship lift is now being used for the outer harbour of the new lock. In order to minimize settlements and to protect the clay seal underneath the harbour basin, vibro compaction has been executed. Results of the pre-construction test-program as well as results of pre- and post-compaction testing are presented in this paper.

1 Project Description

1.1 General Situation

Soon after the reunification of Germany it was decided to expand the canal connection between Hannover, Magdeburg and Berlin, creating an economical, ecological and safe transport system alternative to the crowded highways and

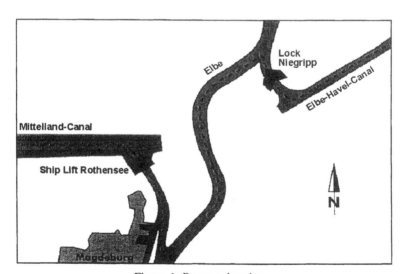

Figure 1: Present situation

motorways. The east-west canal route through the plains of Northern Germany is an important link for the European canal system. Dortmund-Ems-Canal, Mittelland-Canal and Elbe-Havel-Canal connect the industrial districts of the Ruhr Region with Berlin and the River Oder into Poland.

The use of the canal system was severely restricted since its opening 1938, because the crossing of the Elbe was missing. Ships from the west have to leave the Mittelland-Canal through the Ship Lift Rothensee, go south towards the harbour at Magdeburg, then turn north into the Elbe and enter the Elbe-Havel-Canal through the Niegripp Lock. This route is time-consuming and euro-sized barge-convoys cannot use it. Furthermore, low water in the Elbe allows only low capacity ships for several months each year. The present situation is shown in figure 1.

1.2   Previous Construction

As early as 1934 construction started for a canalbridge across the Elbe and a ship lift structure with four 80 meter deep float shafts at the same location as the present construction site. The complicated project progressed well until it had to be terminated in 1942 due to World War II. A major part of the previous project was executed by the Deutsche Schachtbau AG, a company which belongs today to the Bauer Group under the name of Schachtbau Nordhausen GmbH. In the 1970s the area of the ship lift including the deep float shafts have been loosely filled with locally available sand. The top view of the original lock structures is shown in figure 2, a section in figure 3.

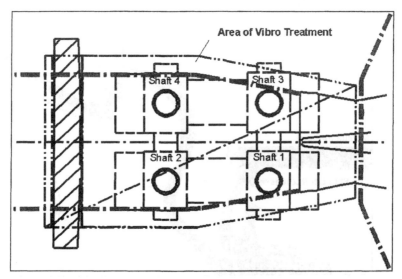

Figure 2: Top view of original lock structure with outline of new upstream outer harbour

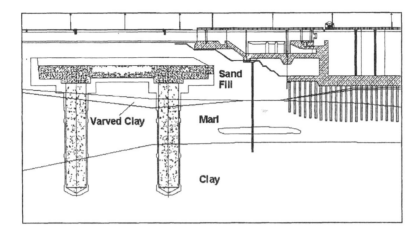

Figure 3: Section of old and new structures in the upstream outer harbour

1.3 New Construction

The new intersection of waterways near Magdeburg consists of the connecting canal system and the following major structures:
- the Elbe Canalbridge (new)
- the Lock Hohenwarthe (new)
- the Lock Rothensee (new)
- the Ship Lift Rothensee (existing)
- the Lock Niegripp (existing)

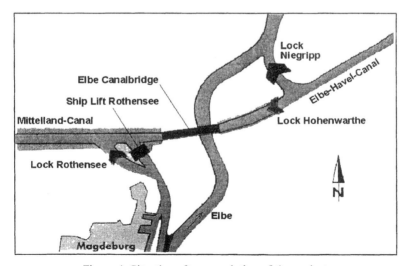

Figure 4: Situation after completion of the project

Completion for the whole project is scheduled for the year 2003 [1]. The situation after completion of the entire project is shown in figure 4.

The Lock Hohenwarthe is a double lock with six water saving thrift basins. The thrift basins save 60 percent of the usual water consumption of the lock operation. The upstream outer harbour of Lock Hohenwarthe is situated directly above the old unfinished structures including the 80 meter deep float shafts, which have been covered with loose sand fill. In order to minimize settlements and to protect the clay seal underneath the harbour basin it was necessary to improve the unsuitable sand fill. Vibro compaction was chosen as an efficient and economical soil improvement method.

2   Design of Vibro Compaction

2.1   Soil Conditions

The sands in the area of the upstream outer harbour have been placed by dumping truckloads onto the half-finished trough and float shaft structures. The material was spread by dozers without any further compaction. Originating from glacial fluvial deposits common to the Elbe valley, the poorly graded medium grained

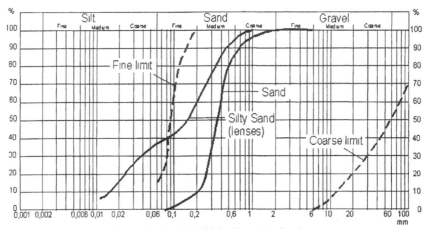

Figure 5: Typical Grain Size Distribution

sand has a coefficient of uniformity $C_u = 2 - 3$. Figure 5 shows the typical grain size distribution with the limits for vibro compaction from [2]. The factor of compactibility $I_f = 0,7 - 0,8$ (1) is very low. Some lenses of silt or clay exist. The typical range of CPT (Cone Penetration Test) values q_s was between 1 MN/m^2 and 8 MN/m^2 and HDP (Heavy Dynamic Probe) values N_{10} between 0 and 10 [3].

$$I_f = (max\ e - min\ e) : min\ e . \tag{1}$$

2.2   Test-Program

Upon recommendation of the contractor a test program was executed, testing three different triangular compaction patterns – 5 m^2 per compaction point (or 2,40 m center-to center distance), 6 m^2 (or 2,60 m) and 7 m^2 (or 2,85 m), see figure 6. Planned treatment depth was 19,00 m or refusal on the underlying original dense layers of gravelly sand or marl. On the average for all test compaction points a treatment depth of 18,00 m could be achieved. Only two compaction points had to be terminated much higher, at 13,00 m depth, because of very dense material, possibly manmade obstructions. The minimum requirement for a safe foundation for the harbour basin was determined to be a CPT value q_s of 6 MN/m^2 or a modulus of elasticity E_s of at least 50 MN/m^2 [4].

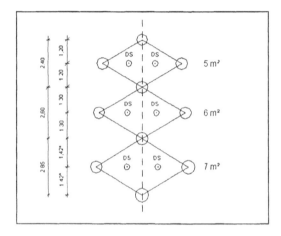

Figure 6: Test Program Layout

2.3   Results and Evaluation

The variation of the test results was relatively high. This was due to the low compactibility, the non-homogenous properties of the sand fill and the existence of small silt lenses. Especially the results of the 7 m^2 grid were inconclusive. It was decided to execute a second test with the 7 m^2 grid. Typical results of pre- and post-compaction cone penetration tests are shown in figure 7. The 7m^2 grid did not fulfill the stipulated density requirements. During testing and evaluation of the results vibro compaction work progressed using a conservative grid of 5 m^2 per point. After evaluation of the test results it was decided to stay with this grid to cover all aspects of the inhomogenous material safely.

The results of vibro compaction tests in tailings of open pit mining in Eastern Germany, a fill material similar to the sands in Hohenwarthe, have been confirmed [5].

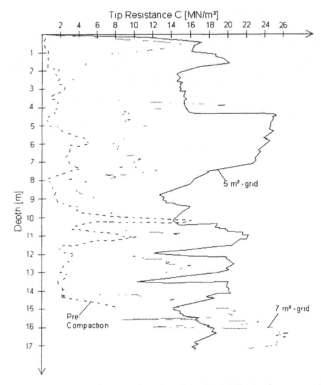

Figure 7: Pre- and Post-Compaction Results (CPT, Tip Resistance only)

3   Project Execution

3.1   Equipment and Method

The project was executed using two crawler cranes Hitachi KH 150 and KH 180 with hydraulic Bauer Deep Vibrators TR 17 and Bauer Power Packs H 180. For technical data see table 1 and table 2

Table 1: Technical Data of Vibrator TR 17

Eccentric Moment	17 Nm
Rotation Speed	3.250 rpm
Centrifugal Force	193 kN
Amplitude at the Tip	+/- 6 mm
Diameter	300 mm

Table 2: Technical Data of Power Pack H 180

Engine Output	118 kW
Hydraulic Pump	180 l/min

Water from groundwater wells was used for flushing and the sand from the lower outer harbour area was added for the compaction process.

3.2 Quantities and Productivity

The penetration depth depended on the distance to the old buried structures, varying from 8,50 m to 30 m. A total number of 1.966 compaction points for a total length of 26.700 lin m were treated. Many treatment columns had to be terminated above the expected depth due to boulders or obstructions. 9.400 m^3 of sand was installed. The average sand consumption was 0,35 m^3/m. The complete work was executed in seven weeks working days and nights. Average productivity was 400 m/rigshift, while the peak productivity was an excellent 617 m/rigshift. Figure 8 gives an overview of the site with two vibro rigs working. In the background the Bauer Trench Cutter BC 30 installs the cutoff-wall down to 53 m.

Figure 8: Site Overview with two Vibro Rigs

3.3   Pre- and Post-Compaction Testing

Typical results of pre- and post-compaction testing are shown in figure 9. All test showed satisfactory results.

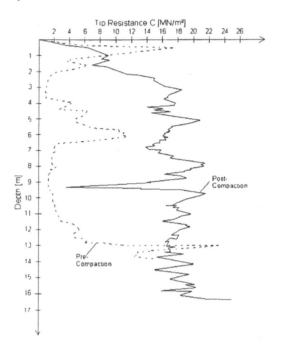

Figure 9: Typical Pre- and Post-Compaction CPT Results

4   Deep Float Shafts

The 80 m deep float shafts were filled with loose to very loose sandy material from top to bottom. The owner decided preliminarily to compact the top 30 m with deep vibrators and to leave the bottom 50 m untreated. Discussions about the risk potential for settlements due to earthquakes are still ongoing. Compaction grouting using a stiff sand-cement-mortar, penetration grouting with micro-fine cement or jet grouting are being considered as possible technical solutions. No decision has been made as of the date of this paper.

5   Summary

The vibro compaction technique has been used successfully to improve the artificial sand fill material on top of a buried and unfinished lock structure for the outer harbour of the new Lock Hohenwarthe, even though obstructions and inhomogeneous material were frequently encountered. High productivity and good cooperation between the owner, Wasserstraßen-Neubauamt Magdeburg, and the contractor, a joint venture between E. Heitkamp GmbH, Herne, Bauer Spezialtiefbau GmbH, Schrobenhausen and Stahlbau Plauen GmbH were the basis for excellent results.

References

[1]   *Wasserstraßenkreuz Magdeburg.* Brochure. Wasserstraßen-Neubauamt Magdeburg (1999).

[2]   Merkblatt für die Untergrundverbesserung durch Tiefenrüttler. Forschungsgesellschaft für das Straßenwesen, Köln (1979).

[3]   *Wasserstraßenkreuz Magdeburg, Vorhäfen Schleusenanlage Hohenwarthe, Baugrundgutachten – 2.Stufe.* Not published. Bundesanstalt für Wasserbau – Außenstelle Berlin (1992).

[4]   *Geotechnische Anforderungen an die Gründung des Projekts Wasserstraßenkreuz Magdeburg – Doppelsparschleuse Hohenwarthe.* Not published. Bundesanstalt für Wasserbau – Außenstelle Berlin (1992).

[5]   GEHRISCH, M. und STEIN, U. (1992): *Bodenmechanische Bewertung von Rütteldruckversuchen in Kippen zur Verhinderung von Setzungsfließrutschungen.* Neue Bergbautechnik, 22.Jg., Heft 8.

2 Verification of compaction

Field tests before and after blasting compaction

Roberto Cudmani and Gerhard Huber

Institute of Soil Mechanics and Rock Mechanics, University of Karlsruhe, Germany

Abstract: Cone penetration tests (CPT) and cone pressuremeter tests (CPMT) are used to investigate the in-situ state of an open pit mine fill before and after single and group blasting compaction. The interpretation of the CPT and CPMT, which is carried out on the base of a hypoplastic cavity expansion model allows an accurate estimation of the density taking into account the influence of depth on the cone resistance and on the pressuremeter limit pressure.

The void ratios determined from CPMT applying the proposed interpretation method agree rather well with the void ratios obtained in the laboratory from "undisturbed" soil specimens sampled by soil freezing. The calculated horizontal earth pressures agree well with the earth pressure measured in the field using spade-shaped pressure cells despite of the strong scattering showed by the latter.

The fill is originally in a extremely loose and rather homogeneous state. The "pressure corrected" relative density remains almost constant and the horizontal earth pressure coefficient decreases slightly with depth. Blasting compaction using single charge blasting appears to be insufficient in order to achieve long term stabilization of the fill since a considerable increase in horizontal stresses is induced but only a small increase of density. Though the effectiveness of the compaction method is slightly increased by detonating charges in groups, repeated blasting seems to be the most successful blasting compaction strategy.

1 Introduction

The Lautsitz region in the eastern part of Germany was a major lignite reservoir (s. figure 1). The lignite deposits with thicknesses of about 10 m lie at depths of about 50 m. The soil covering the deposits consists mainly of fine and medium, well rounded sands of rather uniform grain distribution. The lignite was mined in open pit mines, the main steps being:

- Groundwater lowering up to the bottom of the lignite deposit
- Excavation up to the deposit and mining of the lignite
- Refilling of the mine with formerly overlying sediments
- Restoration of the natural ground water level

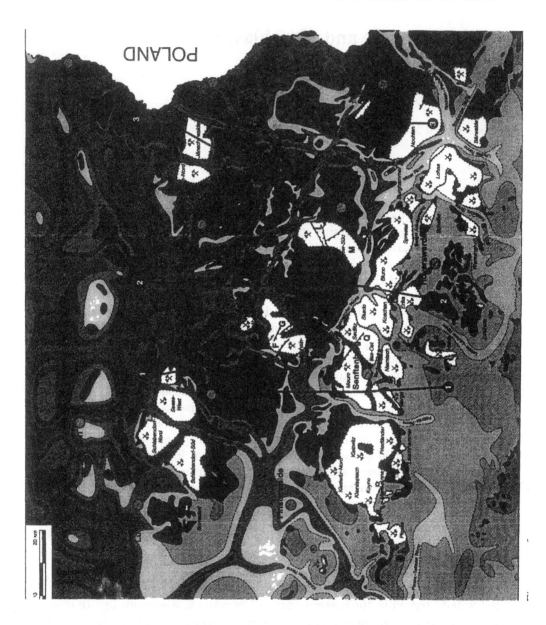

As a result of the filling procedure, the given soil properties and the rising of the ground water table, an extremely loose nearly saturated sand fill is formed with a strong tendency towards spontaneous liquefaction. Typical remediation measures are blasting and deep compaction by means of vibroflotation.

In order to evaluate the liquefaction risk using mathematical models as well as to control the effectiveness of the stabilization measure the state of the fills must be determined before and after remediation. The state of a dry (or fully saturated) cohesionless soils is defined by the density and the stresses as a function of depth. Since the fills are flooded by progressively rising water table gas remains in the skeleton preventing full saturation. Consequently, the degree of saturation is another state variable to be considered.

The density and degree of saturation of the soil can be measured from undisturbed soil specimens sampled by soil freezing. Unfortunately, this procedure is expensive, time consuming and its practical execution is exceedingly difficult. Earth pressure cells can be used to measure the in-situ horizontal stresses, but since the soil is disturbed during the installation of the device the measured value can strongly deviate from the actual value. On the other hand , indirect methods offer a more economical alternative to investigate the state of the soil. However, since the state variables are not measured directly, the accuracy of the approach strongly depends on how accurately the measured soil response can be interpreted.

2   The state variables

As pointed out in the introduction, the state of a simple grain skeleton is determined by the void ratio and the state of stress. If the principal axes of the stress tensor in situ coincide with the vertical and the horizontal directions, the initial stress state is specified by vertical and horizontal stresses, σ_v and σ_h, which are connected by the relation $\sigma_h = K\sigma_v$, where K is the coefficient of earth pressure.

A quantity conventionally used to characterize and compare the states of different soils is the relative density I_D (the density index) defined as

$$I_D = \frac{e_{max} - e}{e_{max} - e_{min}}. \tag{1}$$

where the values of e_{max} and e_{min} are obtained in the laboratory by the use of standard techniques. However, the definition of I_D does not consider pressure dependence of the maximum and minimum void ratios, and hence I_D is not an adequate characteristic of density if a wide pressure range is considered ([6], [18]).

In order to relate the behaviour of a soil in connection with the critical state, BEEN AND JEFFERIES [2] suggested the parameter

$$\psi = e_c - e \tag{2}$$

as a state index. The pressure dependence of the critical void ratio e_c was assumed to be

$$e_c = e_1 - \lambda_c \ln p_0, \tag{3}$$

where $p_0 = (\sigma_v + 2\,\sigma_h)/3$ is the mean pressure, and e_1 and λ_c are constants for a given soil. The parameter ψ was used for the interpretation of CPT ([3], [4]). However, the compression law (3) holds true only in the range of pressures between 0.5 and 2 MPa, depending on the type of soil ([5], [13], [14]). At higher pressures due to the grain crushing, the curve shows a steeper decrease. Since the pressure in the vicinity of the cone tip or a pressuremeter can be rather high, a modification of the function (3) may be necessary. KONRAD [14] used a bilinear logaritmic compression law and the normalized state parameter

$$\psi_N = \frac{\psi}{e_{max} - e_{min}}. \tag{4}$$

This two modifications yield a better correlation of the state parameter with the cone resistance for different soils.

In a similar way we generalize the definition of the density index I_D by using the expression:

$$I_D^* = \frac{e_c - e}{e_c - e_d}, \tag{5}$$

which uses the pressure dependent limit void ratios e_c and e_d instead of constant limit void ratios e_{max} und e_{min}. To describe the dependence of e_c and e_d on the mean pressure p_0 mathematically the function proposed by [1] to model the behaviour of granular material in isotropic compression is adopted:

$$\frac{e_c}{e_{max}} = \frac{e_d}{e_{min}} = \exp - (3\frac{p_0}{h_s})^n \tag{6}$$

where h_s und n are hypoplastic material constants and e_{max} und e_{min} are the usual limit void ratios, which are obtained from equation 6 for $p_0=0$. Typically, if $e_d < e < e_c$, the density index I_D^* lies between 0 and 1, but it can be negative for very loose states if $e_c < e$.

3   Interpretation method

For the interpretation of the cone resistance q_c and the pressuremeter limit pressure p_l obtained from the CPMT the cavity expansion model proposed by [15] will be applied, which uses a hypoplastic constitutive law to simulate mechanical soil behaviour (s. [9] and [19]). Analytical relationships between p_l or q_c and the state variables were derived by CUDMANI [7] and [8] assuming that soil displacements during pressuremeter expansion and cone penetration are similar to those obtained during spherical and cylindrical cavity expansion, respectively:

$$\frac{p_l}{p_r} = \frac{p_{LC}(I_D^*, p_0, K)}{p_r} \tag{7}$$

$$\frac{q_c - p_0}{p_r} = k_q \frac{p_{LS}(I_D^*, p_0) - p_0}{p_r} \tag{8}$$

p_{LC} and p_{LS} are functions describing the dependence of the limit pressures on the state variables for the cylindrical and spherical cavity expansion, respectively. The functions p_{LS} and p_{LC} are determined by computing the limit pressure for different initial states using the cavity expansion model and then fitting the discrete solution with an appropriate function:

$$p_{LC}, \; (p_{LS}) = a(p_0)^b \tag{9}$$

The coefficients a and b are functions of I_D^* of the form: $a = a_1 + a_2/(a_3 + I_D^*)$, $b = b_1 + b_2/(b_3 + I_D^*)$. For a given soil a_i, b_i are constant for the spherical cavity expansion and depend on the earth pressure coefficient K for the cylindrical cavity expansion: $a_i = a_{ii} K + a_{iii}$, $b_i = b_{ii} K + b_{iii}$. All constant and functions a_i und b_i can be unequivocally determined from the numerical solution.

$p_r = 1$ MPa is a reference pressure and k_q is a "shape factor" which takes into account the discrepancy between the actual and theoretical cavity expansion for the case of the cone penetration. The factor was determined by comparing the cone resistance measured in calibration chambers for different sands with the limit pressure p_{LS} obtained from the mathematical model for the same soils and the same initial state ([8]):

$$k_q = 1,5 + \frac{7(I_D^*)^2}{0,15 + (I_D^*)^2} \qquad (10)$$

4   Test field and investigation program

The site chosen for the investigation was located in Kleinkoschen near Senftenberg, state of Brandenburg in the eastern part of Germany and belongs to the former Sedlitz - Skado - Koschen open pit mine complex (s. Figure 2). A typical soil profile of the test field is shown in Figure 3. It consists of three fill layers having a total thickness of about 42 m. The two top layers

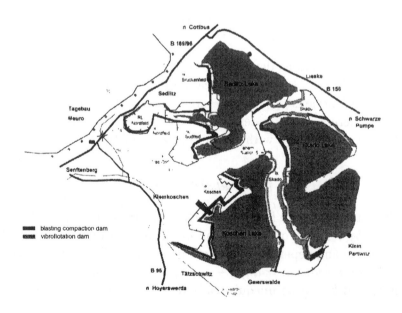

Figure 2: Sedlitz - Skado - Koschen remediation complex

consist mainly on very loosely deposited fine and medium quartz sands with q_c varying between 2.0 to 3.0 MPa. The third layer has some content of fine cohesive material and shows a slow increase of q_c with depth. The three layers are separated by working planes which were compacted during the refilling works by the heavy machinery. The ground water table was at a depth of 5 m.

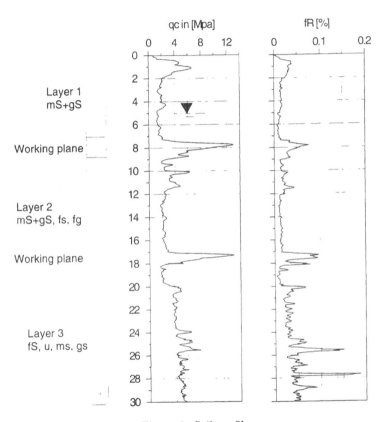

Figure 3: Soil profile

The investigation of blasting compaction was carried out in two stages. In the first stage the compaction effect of single charges was investigated. For that the charges were deployed at the point A, B and C (s. Figure 4). The second stage dealt with compaction produced by group charges. Four tests

(B, D, E, F) were performed, each of group consisting of four blasting points (s. Figure 5). Whereas at B the soil was recompacted with group charges, the groups D, E and F were located in an undisturbed area. A more detailed description of the performed blasting compaction investigation can be found in a companion paper ([12]). CPT and vibropenetration tests were carried out before compaction, after single charge blasting compaction and after group charge compaction. Cross-hole shear wave propagation tests and CPMT were only performed before and after single charge compaction. Only the CPT and CPMT results will be considered in this contribution. The layout of the test fields and the location of the field testing points is shown in Figures 4 and 5 .

5   The cone-pressuremeter test

The cone-pressuremeter is a site investigation device conceived as an alternative approach to pressuremeter testing, in which a pressuremeter probe is incorporated behind a standard 10 or 15 cm^2 cone penetrometer (s. [17]). The pressuremeter has the same diameter as the cone, and is mounted on the penetrometer shaft just above the cone tip (s. Figure 6). Since the insertion of the pressuremeter into the ground is performed as part of the CPT operation, the test is performed out in soil, which has been partially disturbed. However, since the volume of soil affected by a cylindrical cavity expansion is much larger than that affected by the cone penetration its disturbance will be gradually outdone with increasing pressuremeter induced soil deformation, eventually the limit pressure will be mainly controlled by the state of the soil in the far-field.

6   Freeze probing

A sampling campaign using the soil freezing method was carried out previous to the single charge blasting compaction at point B in order to obtain the initial density ([16]). Two cased boreholes with a diameter of 220 mm were drilled down to a depth of 12 and 17 m, respectively. Afterwards a freeze pipe with a diameter of 60 mm was inserted and the casing taken out after filling the gap between freeze pipe and casing with gravel. A frozen soil body with a radius of at least 0.80 m and a height of 3.0 m was produced and samples

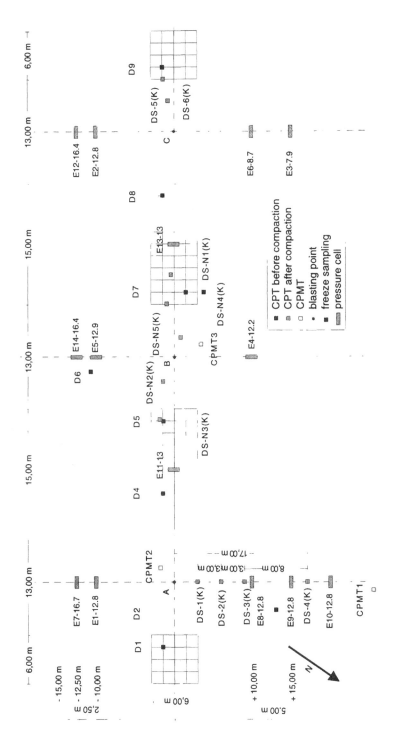

Figure 4: Test field for the investigation of single charge blasting compaction

Figure 5: Test field for the investigation of group charges blasting compaction

Figure 6: cone-pressuremeter device used for performing the CPMT

with a diameter of 0.1 m were retrieved by core-drilling at a distance of 0.6 m from the center of the freeze pipe (s. 7). The distance was regarded as large enough to obtain reliable values of the void ratio.

7 Earth pressure measurements using spade-shaped pressure cells

As part of the program for monitoring blasting compaction 14 push-in spade-shaped pressure cells were installed in the field at various positions and depths to measure changes of earth pressure during and after blasting. The spade cell and the method of installation have been described in detail elsewhere

([10]). It is basically an oil-filled chamber having external dimensions of 150 x
70 x 5 mm (s. Figure 8). The cells were pushed at least 1.0 m into the ground
from the bottom of a borehole, ensuring that the pressure cell was completely
embedded in the soil. After the installation the boreholes were refilled with
gravel.

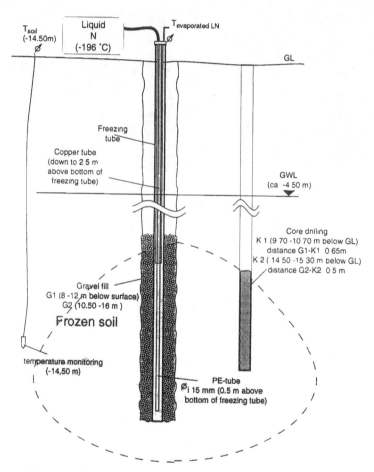

Figure 7: Soil freezing and soil sampling before blasting [16]

Figure 8: spade-shaped pressure cell used for monitoring horizontal earth pressure

8   Results of the CPT and CPMT

Eleven CPT before compaction, 16 CPT after single charge blasting and 8 CPT after group charge blasting up to depths varying between 30 and 40 m were carried out as part of the field testing program to investigate the change of the state due to blasting compaction. Additionally, three CPMT up to depths of 20 m were carried out after the single charge blasting program, each of them including three pressuremeter tests in the depths 5, 10 and 15 m. CPMT1 und CPMT3 were performed at distances of 25 m and 3 m from blasting point A, respectively. CPMT3 was performed at a distance of 3 m from blasting point B.

Figure 9 compares the cone resistances before and after the single charge blastings at 2.0, 5.0, 8.0 and 11.0 m from blasting point B. After single charge blasting q_c increases from 2 MPa up to 6 MPa in the very loose layers below the ground water table and decreases in the work planes, which presumably experiences large deformations when the underlying layers liquefy. The zone affected by the blastings has a radius larger than 11 m (after compaction a settlement trough with a radius of 14 m was measured). That no changes of q_c are observed in the upper soil layer confirms that blasting is not effective above the ground water table.

The improvement of q_c achieved after group charge blasting can be examined with the help of Figure 10, which shows cone penetration diagrams before

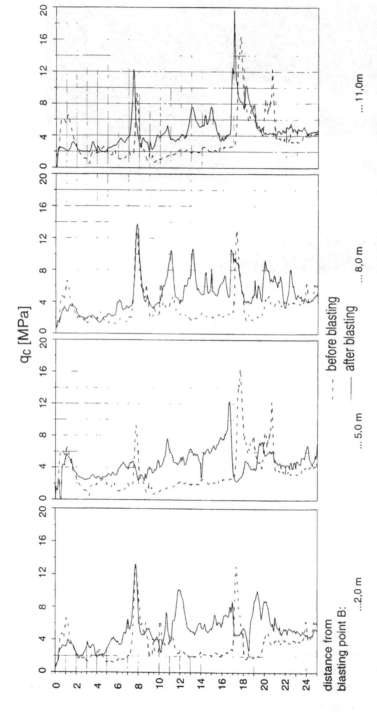

Figure 9: comparison of the cone resistance at different distances from the blasting point B before and after the single charge blasting

and after charge group blasting at D, E and F. The increase of q_c proves only a small improvement in the compaction and thus, in the effectiveness of the compaction method with respect to single charge blasting.

Figure 11 compares the cone resistance measured at the center of charge group B before compaction, after single charge blasting and after group charge blasting. In this case a stronger increase of cone resistance with values up to 30 % higher than those achieved after group charge compaction is observed.

The pressuremeter expansion curves measured in CPMT1, CPMT2 and CPMT3 are plotted in Figure 12. The maximal pressuremeter pressure $p_{max} \approx p_l$ increases after compaction at the three depths considered. The increase of p_l depends on the depth in which the charges are deployed and, like the cone resistance, on the distance from the blasting point. Although site investigation does not showed differences neihter in the material nor in the state at blasting points A und B the increase of p_l is clearly stronger at B than at A.

9   Interpretation of CPT and CPMT

Since granulometric properties in the upper soil layers are similar only one set of hypoplastic parameters was considered for the numerical modelling of the cavity expansion. The hypoplastic parameters were evaluated using the procedure outlined by HERLE [11]. The granulometric properties and the hypoplastic parameters of the fill material, so called "Koschen sand", are listed in Tables 1 and 2, respectively. The coefficients a_l, b_l, a_{ll}, b_{ll}, a_{lll} and b_{lll} for Koschen sand are shown in Tables 3 and 4.

Table 1: granulometric properties of Koschen sand

grain mineral	grain shape	d50 [mm]	U	γ_s [g/cm3]	e_{max}	e_{min}
quartz	rounded	0.50	3.10	2.638	0.44	0.91

Table 2: hypoplastic parameters for Koschen sand

h_s MPa	n	e_{c0}	e_{d0}	e_{i0}	φ_c	α	β
7450	0.11	0.90	0.45	1.04	34	0.14	1.0

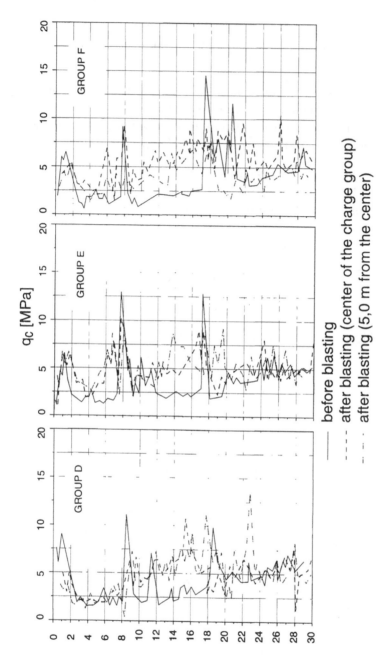

Figure 10: Comparison of the cone resistance before and after group charge blastings D, E and F

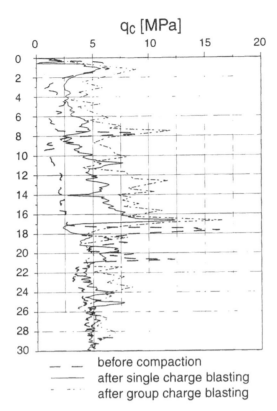

Figure 11: Comparison of the cone resistance before compaction, after single charge blasting and after group charge blasting at blasting point B

Table 3: coefficients a_i and b_i for Koschen sand

a_1	a_2	a_3	b_1	b_2	b_3
1.464	-8.781	-1.201	0.982	0.027	-1.319

Table 4: coefficients a_{ii}, a_{iii}, b_{ii} and b_{iii} for Koschen sand

a_{11}	a_{22}	a_{33}	a_{111}	a_{222}	a_{333}
0.898	-2.265	0.080	1.406	-2.398	-1.372
b_{11}	b_{22}	b_{33}	b_{111}	b_{222}	b_{333}
-0.007	-0.009	0.257	0.985	0.023	-1.611

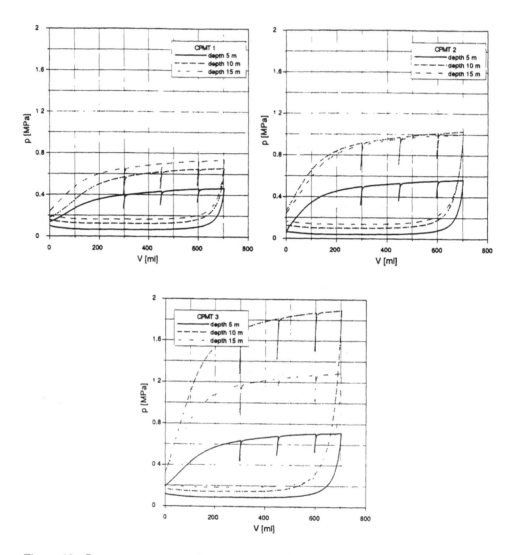

Figure 12: Pressuremeter expansion curves measured in the tests CPMT1, CPMT2 and CPMT3

Figure 13 shows the calculated limit pressures p_{LS} and p_{LC} as a function of the initial mean pressure and the initial relative density I_D^* for Koschen Sand. As it can be seen, the dependence of p_{LS} and p_{LC} on the initial state is qualitatively the same as observed in calibration chambers.

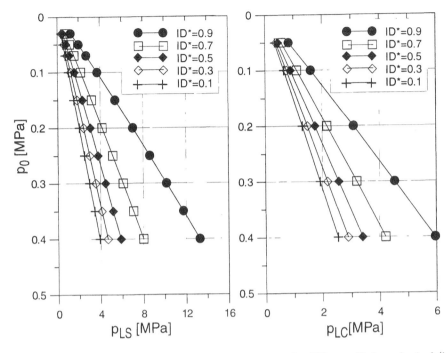

Figure 13: limit pressure as function of the mean pressure for different I_D^* for spherical (left) and cylindrical cavity expansions (right) in Koschen sand

The void ratios and the earth pressure coefficients obtained from the CPMT at depths of 5.0, 10.0 and 15.0 m are presented in Figure 14. The void ratios obtained in the laboratory from the "undisturbed"' soil specimens sampled by soil freezing and the earth pressure coefficients obtained in the field using the spade-shaped pressure cells are included in the diagrams for comparison. Both the density and the earth pressure coefficient estimated from the CPMT agree rather well with the values obtained by soil freezing and earth pressure measurements, respectively. Before compaction I_D^* remains almost constant and the horizontal earth pressure coefficient decreases with depth. The values of K are higher than that calculated after Jaky $K_0 = 1-\sin\varphi \approx 0.45$. Blasting compaction using single charge blasting induces a strong increase of the horizontal earth pressure with values of K up to 1.5 but only a small increase of the density (I_D^* increases in the very loose layers from 0.1 up to a maximal of 0.3). Therefore, single charge blasting does not seem to be very effective in order to achieve long term stabilization of the fill since the horizontal stresses may "relax" if the material is cyclically or dynamically loaded.

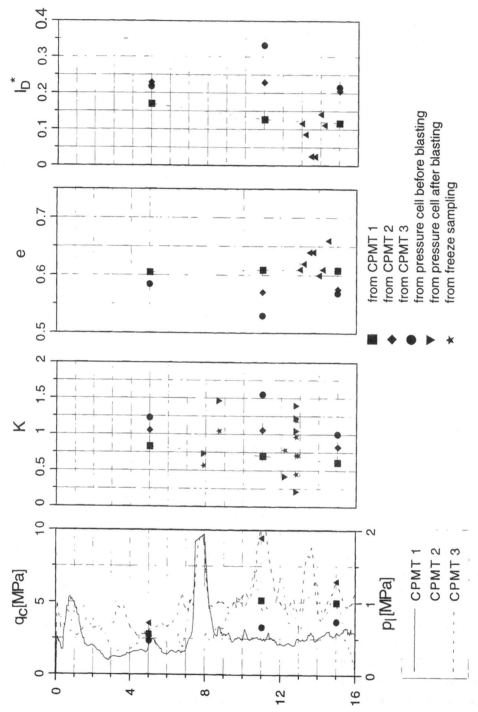

Figure 14: Interpretation of the CPMT before compaction and after single charge blasting

Finally, Figure 15 shows the interpretation of the CPT at point B before and after the compaction. In order to calculated I_D^* the distributions of earth pressure with depth obtained from the CPMT before and after compaction were respectively adopted. Repeated blasting induces an increase of I_D^* up to 0.5 and seems to be the most successful blasting compaction strategy.

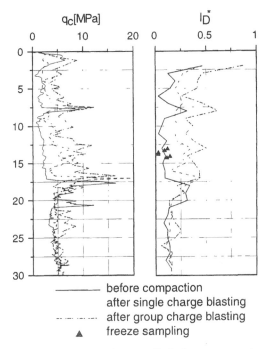

Figure 15: Comparison of the relative density I_D^* before compaction, after single charge blasting and after group charge blasting at blasting point B

10 Conclusions

The in-situ state of open pit mine fills could be realistically estimated using the proposed interpretation method. Blasting compaction induces in general an increase of both horizontal stress and density. However, since the horizontal stress can be reduced by relaxation it is suggested to consider only the improvement of density for controlling the success of the compaction. Single charge blasting was less effective than group charge blasting for increasing the density. On the other hand, repeated blasting (re-compaction) was the most successful blasting compaction strategy.

References

[1] BAUER, E. (1996): *Calibration of a comprehensive constitutive equation.* Soils and Foundation, 36, 1, 13-25.

[2] BEEN K & JEFFERIES, M. G. (1985): *A state parameter for sands.* Géotechnique, 35, 2, 99-112.

[3] BEEN K., CROOKS, J., BECKER, D & JEFFERIES, M. G. (1986): *The cone penetration test in sands: part I, state parameter interpretation.* Géotechnique, 36, 1, 123-132.

[4] BEEN K., JEFFERIES, M. G. CROOKS & ROTHENBURG (1987): *The cone penetration test in sands: part II, general inference of state.* Géotechnique, 37, 3, 285-299.

[5] BEEN K., HACHEY, J. & JEFFERIES, M. G.(1991): *The critical state of sands.* Géotechnique, 41, 3, 365-381.

[6] BEEN, K. (1995): *Cone factors in sand.* Proceedings 1. International Symposium on cone penetration testing. Linköping, 1995.

[7] CUDMANI, R (1996): *Anwendung der Hypoplastizität zur Interpretation von Drucksondierwiderständen in nichtbindigen Böden.* Geotechnik 4.

[8] CUDMANI, R. & OSSINOV V. (2000): *The cavity expansion problem for the Interpretation of cone penetration and pressuremeter tests.* (In preparation)

[9] GUDEHUS, G.(1996): *A comprehensive constitutive equation for granular materials.* Soils and Foundations, 36, 1, 1-12.

[10] GUDEHUS, G. (1999): *BMBF Forschungsprojekt 02WB9520/4: Sanierung und Sicherung fließgefährdeter Kippen und Kippenböschungen.* Final report.

[11] HERLE, I (1997). *Hypoplastizität und Granulometrie einfacher Korngerüste.* Veröffentlichung des Institutes für Bodenmechanik und Felsmechanik der Universität Fridericiana in Karlsruhe, Heft 142.

[12] HUBER, G, CUDMANI, R.(2000): *Experimental investigation of blasting compaction of cohesionless pit mine fills.* Proceedings: International Worshop on Compaction of soils, Granulates and Powders. Innsbruck, Feb. 2000.

[13] ISHIHARA, K.(1993): *Thirty third rankine lecture: Liquefaction and flow failure during earthquakes.* Géotechnique, 43, 3, 349-416.

[14] KONRAD, J. M. (1998): *Sand state from cone penetrometer tests: a framework considering grain crushing stress.* Géotechnique, 48, 2, 201-215.

[15] OSSINOV, V. & CUDMANI, R. (2000): *Theoretical investigation of the cavity problem based on a hypoplasticity model.* (in preparation)

[16] PRALLE N., BAHNER, M. L. & BENKLER, J. (2000): *Computer tomographic analysis of indisturbed samples of loose sands reveal large gas inclusions in the phreatic zone.* (In preparation).

[17] HUGHES, J. M. O. & ROBERTSON, P. K. (1985): *Full displacement pressuremeter testing in sand.* Canadian Geotechnical Journal, 22, 298-307.

[18] SLADEN, J. A.(1989): *Problems with the interpretation of sand state from the cone penetration test.* Géotechnique, 39, 2, 323-332.

[19] VON WOLFFERSDORFF, P.-A. (1996): *A hypoplastic relation for granular materials with a predefined limit state surface.* Mechanics of Cohesive-Frictional Materials, 1:251-271.

Dynamic testing methods for the stiffness of compacted pavements

Enrico Maranini[1] , Giovanni Masé[1] , Gabriele Andrighetti[2] , Alceste Zecchi[2] ,
Vincenzo Fioravante[3]

[1] Department of Engineering, University of Ferrara, Italy,
[2] Amministrazione Provinciale of Ferrara, Italy,
[3] ISMES SpA, Seriate (BG), Italy

Abstract: The load supporting capacity of compacted soils during road construction is generally estimated by static plate load tests. Despite the wide diffusion of this method, its precision and accuracy has not been yet determined. In this work we evaluated the stiffness of stabilised soils during the construction of a road in Northern Italy through the application of two non-destructive methodologies based on acoustic wave propagation. First, Spectral Analysis Surface Waves were used in situ, monitoring the surface vibration induced by a load impact on the ground. Secondly, the propagation of shear acoustic waves was analysed in laboratory, using piezoelectric transducers specially arranged on representative cylindrical samples of soil, tested in triaxial cell. Preliminary results show that these procedures can be successfully applied to control the homogeneity and bearing capacity of bases, sub-bases and wearing surfaces of road paving structures, also as concerns their durability over time.

1 Introduction

During the design of roads and railways the use of local materials for the construction of paving structures may be preferred, with a view to decreasing the construction costs. If the local materials are remarkably sensitive regarding weathering, or they do not have sufficient shearing strength to withstand the stresses imposed by traffic loads, an admixture of cementing agents such as portland cement, lime and hydro-carbonated binders can be used for effective stabilisation, preceding mechanical compaction [1, 2]. In the last few decades this technology has become widely diffused, considering the problems of environmental impact and conservation.

The properties required in road design are commonly defined through static tests such as the Proctor compaction test, CBR bearing capacity and/or the peak strength resulting from uniaxial compression tests (see ASTM standards). Subsequently, during construction, a systematic control of compaction characteristics, flexibility and supporting capacity of paving foundations becomes necessary in order to verify the project requirements. Usual standard procedures recommend the evaluation of humidity, unit weight of soil in place and the bearing capacity resulting from static load plate test, CBR field test or the Benkelmann deflection test. However, uncertainties may exist concerning the results of these methods, especially when applied to thin multi-layered foundations with different characteristics; their application is very limited also as regards durability and the control of stiffness over time.

Dynamic test methods can be used to overcome such problems: as emerges from the theory of elasticity, the propagation of compression waves (P) and shear waves (S) allows the assessment of the constrained modulus M_0 and shear modulus G_0 at very small strains where, as a first approximation, soils behave as linear elastic materials; In particular, the determination of the (dynamic) shear modulus is very useful for characterising the soil during various loading methods. During the last 15 years, acoustic body wave propagation has been widely applied in geotechnical engineering to study states of anisotropic stress, and to determine the coefficient of earth pressure at rest (K_0) of non cohesive soils [3, 4, 5, 6].

On the other hand, each impulse noise induced on a solid surface produces a complex system of vibrations, including P, S and Rayleigh waves (R). In certain conditions, also the Rayleigh waves give a significant contribution to the motion, and their description can be of primary importance for ground characterisation: the velocity of a Rayleigh wave is very close to the shear wave velocity, thus the former can be related to the shear modulus of the soil.

In this work we adopted both approaches to characterise the stiffness of cementing stabilised soils during road construction:

- Spectral Analysis of Surface Waves (SASW) was used to check the stiffness of stabilised pavements. SASW is a non-destructive technique for investigating the rigidity profile of pavements and soil deposits, whose diffusion has widely increased during the last few decades [7, 8, 9, 10, 11]. The methodology is based on the in situ monitoring of Rayleigh wave propagation induced by an impact on the ground.

- Acoustic body waves were transmitted through specimens tested in laboratory conditions, to detect their stiffness properties and state of anisotropy. The wave velocities were related to specimen composition, density and state of stress; this methodology can therefore give detailed information regarding the stiffness characteristics (i.e. G_0) of stabilised soils, both for specimens assembled in laboratory and those cored from the site testing points.

The preliminary results are shown for both test methods; an attempt has been made to compare the results of different methodologies. We have pointed out the main problems encountered during the work. Despite many uncertainties and the necessity for more extensive analysis, we believe that these non-destructive dynamic procedures can be successfully applied for a complete characterisation of the rigidity profile of pavings, during both structure design and road performance verification after construction.

2 Site and soil characterisation by standard test methods

The research was performed in the County of Ferrara, in north-east Italy, where a system of by-pass roads, for a total length of 20 km, is presently under planning and construction. The area is located in the alluvial plain of the River Po: superficial soils are classified as A7-6, A6 and A4, following the AASHO classification system.

The testing sites were chosen along 5 km of road under construction. The structure was designed in embankment, a schematic section of which is shown in figure 1. The subgrade and sub-base courses, for a total depth of 80 cm, were realised using local natural soils, stabilised with lime and/or cement depending on the soil geotechnical properties: admixture of 3% lime or 3%+3% lime+cement was chosen for highly cohesive soils, while the addition of 3% cement was selected for A4 soil types.

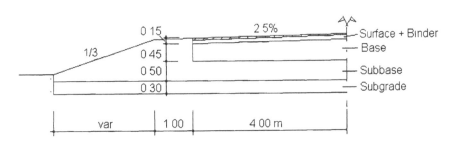

Figure 1: Schematic session of the road embankment

Four representative sections were selected during this work. We divided them in two classes: C1-C2 (whose distances were less than 200 m from each other) and F1-F2. The main differences between the two groups regarded the cementing admixture (3% lime for C1-C2 sections; 3%+3% lime+cement for F1-F2 sections) and the compaction reference methods (modified Proctor compaction for C1-C2; standard Proctor compaction for F1-F2). Another distinction concerned the protocol design for the dynamic field tests, as will be explained in the next section.

Soil characteristics and compaction results obtained in laboratory conditions are summarised in table 1 and table 2: as described in literature, lime addition produced a marked leanness of the plasticity of the clayey soil (liquid limit and plastic index); further mixing with pozzolanic components led to a slight increase of the optimum compaction density when compared with the natural soil.

Table 3 summarises the tests performed during the embankment construction. The stiffness was controlled by static load plate deflection, performed every 250 m of road length using a ϕ30 cm disk plate; each compacted level, 25-30 cm in width,

Table 1: Properties of natural soils

Sect.	Soil	L.L. [%]	P.I. [%]	γ_d opt. [KN/m³]	w opt. [%]
C1-C2	A6	38.9	16.3	18.8 (mod.)	13.0
F1-F2	A6	37.0	16.1	17.0 (std.)	19.0

Table 2: Properties of stabilised soils

Sect.	Mixture	L.L. [%]	P.I. [%]	γ_d opt. [KN/m³]	w opt. [%]	CBR% air/water
C1-C2	3 % CaO	53.5	15.0	17.8	14.0	101.1 / 82.7
F1-F2	3 % + 3% CaO	42.0	11.7	17.4	18.5	51.1 / 40.1

was tested starting from the subgrade course 7-10 days after stabilisation. As expected, the modulus M_d resulting from these tests continuously increased with the thickness of compacted material; no noticeable variations were found when testing the different sections. Uniaxial compression tests were performed on undisturbed cylindrical samples ($L/D = 10/5$ cm) cored from the site testing sections; the uniaxial compressive strength (UCS) showed some scattering, especially as concernead sections C1-C2. From these tests we calculated the Young moduli and the Poisson coefficient: values of $E = 50\text{-}70$ MPa and $\nu = 0.2$ could be considered representative for all the specimens.

Finally, assuming a linear elastic behaviour of the soil, we calculated the (static) shear moduli G for each compacted layer, using the relation

$$G = 0.785\, M_d \frac{1 - \nu}{2 + 2\nu} \tag{1}$$

Also in this case, as observed for M_d modulus, G-magnitude increased with the thickness of compacted material; however, sections C1-C2 were characterised by greater values of G as a consequence of the higher compaction level.

3 Dynamic test methods

3.1 SASW characterisation

The Spectral Analysis of Surface Waves is a non-destructive technique introduced in the early 1980s, and has improved significantly with the development of digital dynamic signal analysers; detailed information regarding the theory of SASW can be obtained in literature [8, 9, 11]. Following this technique, an impulsive noise load

Table 3: Results of the standard control tests

Sect.	Layer	width [cm]	γ_d [KN/m^3]	w [%]	M_d [MPa]	UCS [KPa]	G [MPa]
C1-C2	subgrade	30	14.9	17.9	52	-	14
C1-C2	sub-base 1	25	16.6	14.8	89-70	510(C1)	23-18
C1-C2	sub-base 2	25	15.6	16.0	125-125	410(C2)	33-33
F1-F2	subgrade	30	15.6	15.4	39	-	10
F1-F2	sub-base 1	25	15.3	15.2	53-62	425	14-16
F1-F2	sub-base 2	25	15.3	13.4	80-85	460	21-22

was applied to the surface of the investigation site; two vertical receivers located on the surface, at known distances, were used to monitor the wavetrain generated by the source; the electrical signals were recorded by a digital analyser, and the time signals recorded were transformed to the frequency domain using a fast Fourier transform algorithm. Figure 2 shows the typical experimental set-up, and the equipment used for the acquisition and analysis of the signals.

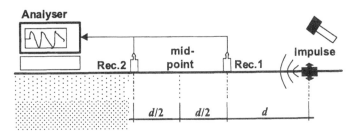

Figure 2: Experimental set-up of SASW field tests

In applying this technique, only Rayleigh waves are taken in consideration. In reality also body waves are produced along the surface; however, as the amplitude of these waves decreases in proportion to $1/r^2$ (r = distance from the source) and the amplitude of the Rayleigh waves decreases in proportion to $1/r^{0.5}$, the former assumption (Rayleigh waves alone) is acceptable for long distances, i.e. if the wavelength considered during the analysis is equal to or more than half of the distance between receivers [8]. Note that a longer wavelength is synonymous of deeper soil investigation; thus, the distances can be successively increased to characterise deeper deposits. In our study, we chose the following protocol for the relation impulse-receiver1-fixed mid point-receiver2 (see figure 2): 2-1-1; 4-2-2; 8-4-4 (distances in meters). Each noise applied to the ground surface contained a wide range of frequencies, thus a great deal of detailed information could be obtained simultaneously.

After calculating, for each frequency f, the phase difference $\phi(f)$ between the two signals, the number of cycles were given by:

$$N(f) = \frac{\phi(f)}{2\pi} \ . \tag{2}$$

Since the distance d between the receivers was known, we could calculate the wavelength λ ($\lambda = d/N$) for each harmonic; hence the velocity (phase velocity) was defined as:

$$V(f) = \lambda f = \frac{d}{N} f = \frac{2\pi(f)d}{\phi(f)} \ . \tag{3}$$

Figure 3 shows a typical dispersion curves obtained from the testing site point C1.

Once the experimental dispersion curves were defined, an inversion process was necessary to backcalculate the elastic properties for the layers characterising the compacted soil profile. We followed the iterative procedure proposed by NAZARIAN [8], who suggested:

   a) a first attempt to attribute the soil parameters (ρ, ν, H_i) for the multi-layered compacted soil;

   b) comparison between the theoretical dispersion curve and experimental curve obtained in the field;

   c) adjusting of parameters until the best fit between the two curves is obtained.

It is useful to remember that the Rayleigh wave velocity (V_R) is very similar to the shear waves velocity (V_S), thus it reflects the rigidity (G_0) of the deposit, according to:

$$G_0 = \rho V_S^2 \ . \tag{4}$$

During our characterisation we started the inversion process by using the following parameters for both profiles C1-C2 and F1-F2, as suggested from the standard tests described in the previous section: $\rho= 1.9$ Mg/m^3; $\nu= 0.2$.

Figure 4 shows the velocity profiles of sections C1-C2, which were performed few weeks after the sub-base course stabilisation. More exactly, on section C1, only the subgrade course and the first 25-30 cm of the sub-base course were compacted before testing; hence, the total stabilised soil depth was more or less 45-50 cm. On the contrary, on section C2, stabilisation of the sub-base course was completed; the embankment depth was around 80 cm, including the subgrade layer (30+25+25 cm). A third SASW profile was performed on the natural soil deposit, just besides the road sections, but not compacted or mixed with lime.

Similar results obtained on the sections F1-F2 are shown in figure 5: in this case the dynamic tests were performed after road completion, directly on the wearing

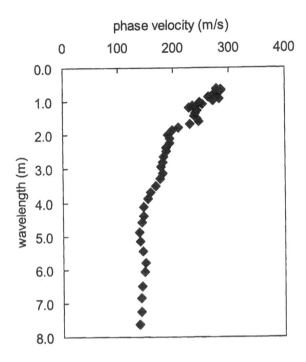

Figure 3: Dispersion curve obtained from the testing site point C1

surface. The entire thickness of the embankment (approximately 140 cm) could be investigated. Again, a third profile is shown, regarding a SASW analysis performed on a road section very close to the testing site points, whose embankment was built without any admixture of cementing material.

The table 4 and table 5 summarise the results of the previous diagrams in terms of stiffness, i.e. shear modulus G_0.

From all these schemes, the advantages provided by SASW characterisation are quite clear: each single layer of the road could be identified, and a direct response was given concerning soil stiffness and compaction, independently of the depth of the layer from the testing surface. For example, profile C2 showed a weakness in the first 40 cm, which was probably due to some problems occurring during mixing or compaction of the first layer. Similar conclusions can be drown for sections F1; furthermore, in this case, the characterisation of the entire embankment was clearly evident in spite of the wearing surface, whose stiffness was several times higher than the underlying layers.

In all cases the velocity profiles of the natural soils were well differentiated from those resulting from the stabilised soils.

Table 4: Rigidity profiles for sections C1-C2

Sect.	Layer	elevation H_i [m]	G_0 [MPa]
C1	sub-base 1	0.30-0.55	105
C1	subgrade	0.00-0.30	170
C2	sub-base 2	0.55-0.80	80
C2	sub-base 1	0.30-0.55	105
Natural	-	0.00-1.00	35

Table 5: Rigidity profiles for sections F1-F2

Sect.	Layer	elevation H_i [m]	G_0 [MPa]
F1 -F2	binder+surf.	1.25-1.40	> 500
F1 -F2	base	0.80-1.25	230
F2	sub-base	0.30-0.80	120
F2	subgrade 1	0.00-0.30	80
F1	base 1	0.80-1.25	90
F1	sub-base 1	0.30-0.80	80
F1	subgrade 1	0.00-0.30	80

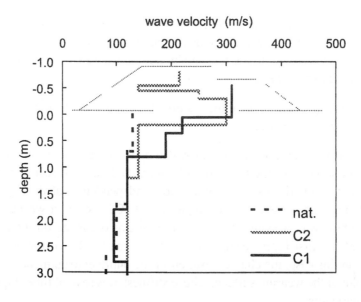

Figure 4: Comparison of velocity profiles of SASW performed on sections C1-C2 and natural soil

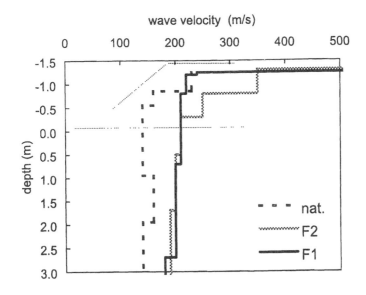

Figure 5: Comparison of velocity profiles of SASW performed on sections F1-F2 and natural soil

3.2 Body wave propagation in laboratory specimens

Over the past 20 years, laboratory determination of the dynamic deformational characteristics of soils have been conducted using large equipment including calibration chambers and big specially assembled cylindrical/cubic samples [3, 4, 5, 6]. The system of Acoustic body wave propagation presented here is a technique recently introduced to evaluate the deformational characteristics of soils using a very common apparatus: the triaxial cell [12, 13, 14]. Following this method, both the costs and difficulties of performing such tests could be heavily reduced. The measuring technique involves the use of small piezoelectric transducers specially arranged on cylindrical specimens [14]; the number of transducers and their orientation is related to the state of stress and the intrinsic heterogeneity of the tested material. As emerges from the theory of elasticity, a complete characterisation of stress-strain relations for a transversely isotropic medium requires the determination of five independent elastic constants; consequently, a corresponding number of couples of transducers should be assembled considering the direction of axial-symmetry, as shown in figure 6.

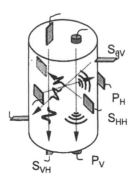

Figure 6: Scheme of transducers arrangements for a transversely isotropic medium

During this study we confined our interest to the definition of the dynamic shear moduli of some specimens prepared in such a way as to represent the actual road section; the mechanism of compaction produced, in the soil deposit, a stress-induced anisotropy approximated by that of a transversely isotropic medium with a vertical axis of symmetry. Consequently, two types of shear waves were propagated inside our specimens:

- S_{VH}, propagating in vertical direction with the particles vibrating in horizontal direction;

- S_{HH}, propagating in horizontal direction with the particles vibrating along the horizontal plane.

The tests were performed by transmitting a series of single sinusoidal pulses with a frequency ranging from 5 to 30 kHz. A function generator and a power amplifier were used to drive the pulses; the received signal was conditioned by an amplifier, then analysed by an oscilloscope in the time domain and digitally recorded. Figure 7 shows the typical experimental set-up.

As the sample dimensions L/D were known, the shear wave velocities were calculated by $V_{VH} = L/t$ and $V_{HH} = D/t$. The shear moduli G_V and G_H were thus found using (4), for both the directions of symmetry.

Two types of samples were tested; the first group came directly from the field, cored near the SASW testing points F1-F2 and C1-C2. Body waves were propagated on specimens with dimensions D/L= 5/10 cm; during testing, we applied a law isotropic pressure of 20-50 KPa to reproduce the field stress conditions. The results

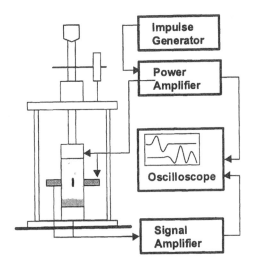

Figure 7: Experimental set-up for the acoustic wave propagation during a triaxial test

Table 6: Dynamic laboratory tests performed on triaxial specimens cored from site

Sect.	Layer	ρ [Mg/m^3]	V_{VH}/V_{HH} [m/s]	G_V/G_H [MPa]	ratio
C1	sub-base 1	1.9	459 / 388	395 / 286	1.38
C2	sub-base 2	1.9	426 / 404	344 / 310	1.11
F1	sub-base	1.9	376 / -	269 / -	-
F2	sub-base	1.9	365 / 275	253 / 143	1.77

are shown in table 6. We point out that the same specimens were then tested in uni-axial compression, the results of which have been briefly discussed in section 2 (see table 3).

The second group of samples were specifically assembled in laboratory, using soil from sections C1-C2, and compacted following the modified Proctor compaction method. Specimens of both natural soil and soil stabilised with 3% CaO were tested, at different water contents, to detect the influence of lime addition on the dynamic stiffness magnitude (see table 7).

Looking at the results of this section, the state of anisotropy due to mechanical compaction characterising all the specimens is particularly evident; however, ratios G_V/G_H vary widely from 1.01 to 1.7, without any trend. As expected, compaction performed in laboratory conditions produced apparently higher stiffness with respect to what was observed on the field; this is also confirmed by the average densities of the tested specimens.

Table 7: Dynamic laboratory tests performed on compacted specimens prepared in laboratory

Sect.	Stabilis.	ρ [Mg/m^3]	w [%]	V_{VH}/V_{HH} [m/s]	G_V/G_H [MPa]	ratio
C1-C2	natural	2.1	12	494 / -	512 / -	-
C1-C2	natural	2.1	14	457 / 451	438 / 427	1.03
C1-C2	natural	2.1	16	378 / 298	330 / 228	145
C1-C2	3 % CaO	2.1	12	634 / 600	844 / 756	1.12
C1-C2	3 % CaO	2.1	14	567 / 537	675 / 606	1.12
C1-C2	3 % CaO	2.1	16	533 / 529	596 / 588	1.01

It is interesting to note that the wave velocities reported in table 7, and consequently the shear moduli, tend to decrease in magnitude with higher contents of mixing water, whether if the water content was above or below the optimum value (see reference values in tables 1 and 2).

4 Discussion and Conclusions

A direct comparison of the results of the tests obtained following the different procedures is quite problematic. We have to remember that the propagation of waves through soil deposits allows an assessment of the deformation moduli at very small strains (less then 0.001%) where, as a first approximation, soils behave as linear elastic materials. On the contrary, standard procedures aimed at the bearing capacity of soils give values which are representative for greater deformations (0.2-0.3%). It is well known that the soil stiffness decrease monotonically as the strain increases; but a relation between strain level and stiffness magnitude is not so immediate, depending on scale effects, state of internal stresses and boundary conditions.

The Static load plate test gave results with good repeatability and easy adaptability (table 3). On the other hand, it did not permit to verify accurately the quality of construction: the method was not predictive of the real rigidity profile of the embankment, because each compacted layer was too limited in thickness. The depth involved by the pressure-bulbs resulting from the plate deflection is generally greater than the plate diameter; thus, the stiffness resulting from the load test performed on the road surface was heavily influenced by the response of the underlying layers.

On the opposite, informations regarding the stiffness profile of the multi-layered deposit were widely pointed out by the SASW characterisation, which revealed each single discontinuity concerning the rigidity profile. During this work the Seismic Analysis of Surface Waves gave excellent results. SASW is a non-destructive method characterised by a relative economy, rapidity of execution and good repeatability of the stiffness profiles. Some difficulties can occur during the inversion process of the experimental dispersion curve, depending on the type of deposit and stratification; however, such methodology allows the analysis of several meters of soil below the ground surface, adapting the geometry of receivers and impulse load.

Shear wave velocities determined by both SASW and body wave propagation exhibited meaningful differences, due to scale effects and boundary conditions. Specimens tested in laboratory and cored from the site testing points (table 6), showed higher shear moduli with respect to what resulted from SASW characterisation (tables 4 and 5), although both methods induced (small) strains of the same entity. This was a probable effect of microcracks distribution, which appeared at the field scale but vanished at the scale of laboratory specimens, and was thus recognisable by means of SASW procedure only. According to this, the last group of specimens, compacted directly in laboratory following usual compaction methods (table 7), showed even higher values of shear stiffness, not comparable to those obtained from the first group of samples.

We remind that the propagation of acoustic body waves inside cylindrical specimens of stabilised soil represents a new application, which requires further investigations. At present, we cannot compare the results with those obtained by Seismic Analysis of Surface Waves method. Nevertheless, the former procedure can be utilised to verify the best performance of soils assembled and compacted in laboratory, in terms of relations between shear stiffness, cementing addition and optimum dry density and water contents. Our final intention is to arrive to a complete characterisation of the road foundations by means of dynamic tests alone, both from laboratory assembling and field control tests. This is is the subject of continuing research and investigation.

Acknowledgements: The Authors wish to thank ISMES and Amministrazione Provinciale of Ferrara for their authorisation to publish the results of this research. Special thanks to Mr. R. Capoferri, of the Geotechnical Division of ISMES for his valuable support during SASW characterisation and useful discussion and advice during the work.

References

[1] KEZDI, A. (1979).*Stabilised Earth Roads* , Elsevier.

[2] NATIONAL LIME ASSOCIATION (1991). *Lime stabilization Construction Manual* , Bullettin 326.

[3] BELLOTTI, R., JAMIOLKOWSKI, M., LO PRESTI, D.C.F., O'NEILL, D.A. (1986): *Anisotropy of small strain stiffness in Ticino sand.* Gotechnique, 46, 1, 115-131.

[4] FIORAVANTE, V., JAMIOLKOWSKI, M., LO PRESTI, D.C.F., MANFREDINI, G., PEDRONI, S. (1998). *Assessment of the coefficient of the earth pressure at rest from shear wave velocity measurements.* Gotechnique, 48, 5, 657-666.

[5] KOHATA, Y., TATSUOKA, F., DONG, J., TEACHAVORASINSKUN, S., MUZUMOTO, K. (1994). *Stress states affecting elastic deformation moduli of geomaterials.* Proc. of Ist Int. Symp. On Prefailure Deformation Characteristics of Geomaterials, IS, Sapporo, vol. 1, 3-10.

[6] STOKOE, K.H. II, LEE, J.N.K., LEE, S.H.H., (1991): *Characterisation of soil in calibration chamber with seismic waves.* Proc. ISOCCT1, Potsdam, N.Y.

[7] NAZARIAN, S., STOKOE, K.H. II, HUTSON, W.R. (1983): *Use of spectral analysis of surface waves for determination of moduli and thickness of pavements systems.* Transportation Reaserch Record, n. 945, Washington D.C.

[8] NAZARIAN, S. (1984): *In situ determination of elastic moduli of soil deposits and pavement systems by Spectral Analysis of Surface Waves method.* Ph.D Thesis, Civil Engineering Dept., University of Texas at Austin.

[9] SANCHEZ-SALINERO, I., ROESSET, J.M., SHAO, K.H., STOKOE, K.H., RIX G.J. (1987): *Analytical evaluation of variable affecting surface wave testing of pavements.* Annual Meeting of the Transportation Research Board.

[10] STOKOE, K.H. II, RIX, G.J., NAZARIAN, S. (1989): *In situ seismic testing with surface waves.* Proc. 12th Int. Conf. on Soil Mechanics and Foundation Engineerings, Rio De Janeiro, 331-334.

[11] LAI, C.G., RIX, G.J., (1998): *Simultaneous inversion of Rayleigh phase velocity and attenuation for near surface site characterisation.* Georgia Institute of Technology.

[12] DYVIK, R., MADSHUS, C., (1985): *Laboratory measurements of Gmax using bender elements.* Proc. Am. Assoc. Civ. Engrs. Convention, Detroit, American Society of Civil Engineers.

[13] VIGGIANI, G., ATKINSON, J.H. (1995): *Interpretation of bender element tests.* Gotechnique, 45, 1, 149-154.

[14] BRIGNOLI, E.G.M., GOTTI, M., STOKOE, K. H. II, (1996): *Measurement of shear waves in laboratory specimens by means of piezoelectrical transducers.* ASTM Geotechnical Testing Journal, GTJODJ, vol.19, 4, 384-397.

Compaction control of a soil improvement with powder columns

Wolfgang Wehr[1] and Joachim Berg[2]

[1] Keller Grundbau GmbH, Overseas Division, Offenbach, Germany
[2] Keller Grundbau GmbH, Development and Consulting Division, Offenbach, Germany

Abstract:

Compaction control has been performed by means of crosshole measurements. The soil consists of an inhomogeneous coal mining deposit. Powder columns consisting of fly ash have been installed as soil improvement by the Keller Grundbau GmbH using a modified Vibroreplacement technique.

Crosshole measurements before and after the column installation have been applied to monitor the improvement of the soil. Shear wave velocities have been measured along three different lines in two depths.

The obtained results are the earth pressure coefficients determined with normalised stresses and the void ratios using an empirical correlation. K0-values before the installation of the powder columns decrease considerably after the installation and void ratios changed from 0.9 to 0.4, showing an impressive soil improvement due to the powder columns.

A comparison of the crosshole results with density measurements from borehole logs and with earth pressure measurements yields a good agreement.

1   Introduction

For all foundation engineering tasks it is helpfull or even necessary to measure the state parameters of a soil. Civil engineers are especially interested in the density or void ratio of the soil and soil stresses which are related to each other with the earth pressure coefficient.

A lot of effort is spent on the recultivation of old coal mining areas in south-east Germany nowadays and various soil improvement methods are tested in order to find the most economic one. In order to test their effectiveness, methods like crosshole measurements using shear waves are applied to measure the above mentioned state parameters.

2   Stratigraphy and installation of powder columns

Test area A2 is situated on an area owned by Lausitzer and Mitteldeutsche Bergbau- verwaltungsgesellschaft mbH (LMBV) near the town of Leipzig. The institute of soil mechanics and rock mechanics of Karlsruhe University investigated the effectiveness of different soil improvement schemes.

The soil consists mainly of sand with a fines content of approximately 20%, although some soil layers exist with a silt content of up to 60%. The soil is by no means homogeneous and does not have horizontal layers, because it has been deposited with conveyer belts forming soil cones.

A test field of 50 x 30 m has been improved with powder columns by the Keller Grundbau GmbH. Within this area 740 powder columns were arranged in a square grid of 1,35 x 1,51 m = 2,04 m^2 per powder column. The planned depth is 7,5 m below working level and the average consumption of powder amounts to 0,36 t/lin m. The loosest, respectively the densest state of the fly ash, which was used as material, was determined as $\rho_l = 1,17$ g/cm^3 and $\rho_d = 1,57$ g/cm^3. As a result the following diameters were determined: loose $d_l = 0,63$ m and dense $d_d = 0,54$ m. According to the observations made, a dense state can be assumed as the columns could be stepped on immediately after production and a sounding rod could only slightly penetrate. Furthermore, excavations have confirmed these values.

The columns did not reach the ground water, but nevertheless, the fly ash developed stiffness with the humidity contained in the soil, even though with a delay.

The columns were installed by using equipment especially designed for the installation of powdered columns. This machinery set consists of a caterpillar vibrocat, system Keller, with a vertical leader to which a vibrator is connected, which can be activated. To activate means that, by using a winch system, a part of the caterpillar weight in addition to weight of the vibrator itself, including equipment and filling, may be surcharged during the sinking process. The depth vibrator is equipped with a lateral tube which transports the powder. The tube aperture is located at the vibrator tip (bottom of the vibrator). Above the vibrator an oval tube is located which, in the plan view, transcribes the vibrator and its lateral tube. The lateral tube at the top of the vibrator branches off from the oval tube. Above the oval tube a container with an air outlet is attached into which the powder is blown tangentially.

The oval tube and the container serve as storage facility in order to control the powder supply and powder consumption. Three indicators are installed which show filling or emptiness of the container, as well as the emptiness of the oval tube. At the transition between the oval tube and the lateral tube a device is attached in order to supply the powder.

The powder is taken out of a silo and is transported pneumatically to the container of the installation device.

Having filled powder into the container and the oval tube the vibrator is started and lowered - if necessary by means of activation.

After having achieved the final depth, the production of the powder column is started by retracting the vibrator. During the retracting process the powder is emerging from the vibrator tip into the open space. When the powder level in the container or in the oval tube is low, the supply from the silo is starting. A well-adjusted team of personnel will be able to achieve a simultaneous supply and consumption of powder.

The construction of powder columns can be achieved by various methods, e.g. by the execution of a shaft by continuous pulling, by repeated lowering after the first production of a shaft including compaction, i.e. by lifting the vibrator to a certain extent and lowering it again. At each step powder is emerging, which is at once compacted and replaced. After a certain amount of compaction steps, the working level, respectively a certain defined depth below working level is reached, and then the vibrator is moved to the next column to be installed.

Controlling this soil improvement with powder columns, crosshole measurements have been carried out before and after the column installation.

3   Principle of shear wave measurement

Shear waves are used for the measurements, because it is possible to exite these waves in a decoupled mode, Heelan [7]. Furthermore the wave signal inverts, if the exitation is also inverted. This enables a clear identification of the measured signals, Haupt [6]

Seismic methods can be distinguished by the relativ position of the wave source and the geophones. Here the crosshole technique is applied where the source and the geophones are situated in boreholes at the same depths. Fig.1 shows the scheme of the crosshole measurements. After the source has been exited the time of the wavefront between two geophones is measured. After a normalization of the stresses the void ratio and the earth pressure coefficient may be evaluated. This is possible using differently orientated wave fronts. Each shear wave posseses a direction of propagation and a direction of polarization, as shown in Fig.2.

There are many possibilities to exite a wavefront. Apart from explosions and periodic exitations which are not frequently used, single impulses with an hammer or shear vane source are applied. Hoar and Stokoe [8] proposed the shear vane source which is connected to a steel rod decoupled from the surrounding soil. With this steel rod it is possible to exite vertically polarized and horizontally propagating waves S_{vh} as well as horizontally polarized and propagating waves S_{hh} depending on the shape of the exiter. In the first case a falling weight is used which may be dropped on or lifted to a steel plate and in the second case hammer blows are used sending impulses to the steel rod perpendicular to to measured path.

Aufzeichnungsgerät

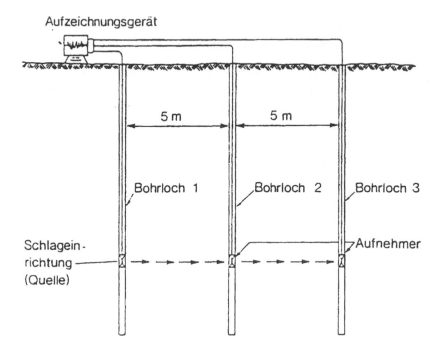

Figure 1: Scheme of crosshole measurements |17|, exiter=Quelle, borehole=Bohrloch, geophone=Aufnehmer

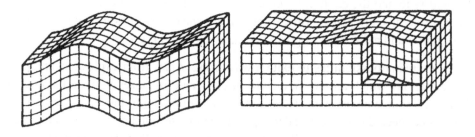

Figure 2: Shear wave with vertical and horizontal direction of polarization, [12]

The difference of the measured times between the same two geophones in both cases is related to the stress distribution in the soil in the measured depth. Evaluation these informations the earth pressure coefficient may be calculated according to Jamiolkowski et al. [10]. Afterwards the void ratio may be found using an empirical correlation including normalized shear wave velocities.

An uncertainty of the measurements is the assumption of straight wave paths in the soil, because it is only feasible to monitor the incoming signals but not the wave paths. According to Behle [1] it has to be taken into account that an elliptical wave propagation in the soil leads to wave splitting on inclined soil layer boundaries. Therefore the evaluation of the signals is easier in homogeneous soils with horizonatal layers.

3.1 Calibration with resonant column tests

Crosshole seismic and resonant column (RC) tests both make use of shear waves. Therefore a correlation between field and laboratory measurements may be established. RC-tests are well known in soil dynamics to estimate the shear modulus G and the shear wave velocity v_s of the soil.

Soil specimen have been taken during the execution of crosshole measurements on a test field in Zwenkau nearby, Schellhase [16]. RC-tests have been performed evaluating v_s depending on the void ratio and the water content of the soil.

Averages of a test series with the water content between 10-12% have been used to establish a linear relationship between the void ratio e and v_s. This relationship is used here for calibration purposes:

$$e = -0.00917 \cdot v_s^n + 2.52 \tag{1}$$

A similar equation has been found for a pit mining area in the Lausitz region, where crosshole measurements have been applied to determine the success of soil compaction by means of blasting, Wehr et al. [18].

4 Layout of field measurements

The test layout of Raju [14], Gudehus [5] and Wehr et al. [18] has been modified to exite horizontally propagating and vertically polarized shear waves as well. Two shear vanes and twelve geophones have been installed in depths of -3m and -6m having three measurement paths in each depth. In this publication only the depth of -3m will be evaluated as an example, Fig.3. Each shear vane has a length of 2m, a diameter of 0.24m and is fixed to a steel rod with which it has been pressed into the soil. The steel rod is decoupled from the soil using a steel tube with a larger diameter. The impulses are applied to the rod just above the ground level.

The shear vanes both have the same distance of 10m from the test field in which the power columns will be installed later. Each of the three measurement paths is equipped with two geophones installed in the depth of the center of the shear vane. All geophone units consists of three single geophones in all the space directions.

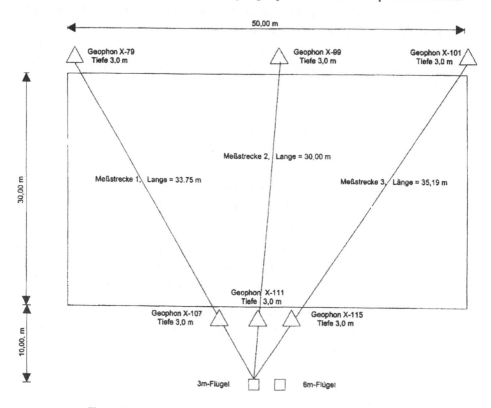

Figure 3: Test field A2 with measurement paths in a depth of 3m. [3]

Two different exitations have been chosen to exite the type of shear waves listed below.

S_{hh} = horizontally polarized and horizontally propagating shear wave with the largest amplitude in the X-direction (90 degrees horizontal to the measurent path), exitation with hammer or alternativly torsion

S_{vh} = vertically polarized and horizontally propagating shear wave with the largest amplitude in the Z-direction (vertically), exitation with falling weight

Exiting the S_{hh}-waves a 5kg sledge hammer with a rubber head is used to execute hammer blows against the steel rod. To invert the signals the hammering direction is changed 180 degrees. Exiting the S_{vh}-waves a 5kg falling cylindrical weight guided on a steel rod is used to execute blows against a steel plate. To invert the signals the weight is pulled up against another steel plate.

Each test consists of 8 similar exitations and measurements of the signals. The average value is used to minimize all time dependent perturbations. All three geophone components (X-,Y-,Z-direction) are recorded to check the polarisation of the waves.

In November 1997 the first series of measurements have been executed before the installation of stone columns. In Mai 1998 a second series of measurements have been performed 60 to 80 days after the installation to allow for the stiffening of the columns.

5   Evaluation of results

Shear velocities are determined using a cross correlation. A comparison of the original signal with the inverted one shows the existance of shear waves and a comparison of the amplitudes of the geophone components displayed the largest amplitudes in the directions of polarisation and propagation and a neglegible amplitude in the third direction.

The shear wave velocities after the installation of the powder columns have increased to a large extent and thus the expected densification of the soil has taken place. To precisise this qualitative statements it is necessary to normalize the velocities with respect to stresses. Afterwards the earth pressure coefficients and the void ratios may be calculated.

5.1   Earth pressure coefficient and stresses

There are many methods to measure stresses in the soil. In contrast to direct methods like pressure cushions or earth pressure spades, where a single measured value is indicated for a certain point, average values over a certain distance may be measured using seismic methods, Blasberg [3].

In 1985 Stokoe and later Jamiolkowski et al. [10] proposed to determine the earth pressure coefficient with differently orientated shear waves according to the following formula.

$$K_0 = C \cdot [v^s_{hh}/v^s_{vh}]^8 \tag{2}$$

with
K_0 = earth pressure coefficient

C = soil constant

v_{hh}^s = shear wave velocity of horizontally polarized and propagating wave

v_{vh}^s = shear wave velocity of vertically polarized and horizontally propagating wave

The soil constant C is the eighth power of a quotient of two soil constants which are denoted with $C_s(hh)$ and $C_s(vh)$. As a physical meaning the degree of the structural soil anisotropy at low strains is assumed. Bellotti et al. [2] confirm this assumption with laboratory tests. In a pressure chamber they measured an average value of $C_s(vh)/C_s(hh)=0.91$ for Tizino sand. Therefore the constant $C=0.47$. Fiorante et al. [4] found similar values for Kenia sand and Jamiolkowski [11] in additional tests as well. Although it has to be pointed out, that this value is not a generally valid constant, but depends on the anisotrophy of the soil. In this case of test field A2 a constant of $C=1.0$ is assumed, because a homoeneously directed anisotrophy is very unlikely.

In the next step the shear wave velocities are normalized with the stresses depending also on the depth. This is necessary to compare the crosshole velocities with the RC-test velocities. The normalisation is performed using the following formula.

$$v_{sn} = v_s[m/s]/(0.5((\Sigma_a + \Sigma_p)[kN/m^2]/p_a[kPa])^n) \tag{3}$$

with

v_{sn} = normalized shear wave velocity

n = empirical exponent 0.25 after Roesler [15]

Σ_a = stress in propagation direction of shear wave

Σ_b = stress in polarization direction of shear wave

p_a = air pressure; 100 kPa

Considering $\sigma_x = \sigma_y = K_0 \cdot \sigma_z$ and $\sigma_z = \gamma h$ yields the following equations for the normalized shear wave velocities:

$$S_{hh} - waves : v_{hh}^n = v_{hh}/(\gamma h K_0/100)^{0.25} \tag{4}$$

$$S_{vh} - waves : v_{vh}^n = v_{vh}/((\gamma h K_0 + \gamma h)/200)^{0.25} \tag{5}$$

with

v_{hh}^n, v_{vh}^n = normalized shear wave velocities

v_{hh}, v_{vh} = measured shear wave velocities

h = measurement depth

K_0 = calculated earth pressure coefficient

5.2   Void ratio

The unit weight γ may either be estimated or be calculated with an iteration process including the void ratio.

In order to start the iteration the water content and the specific unit weight or the soil has to be estimated. Using the common soil mechanical formula

$$e = (1 + w)\gamma_s/\gamma - 1 \qquad (6)$$

the only unknown parameters are e and γ. The void ratio may be substituted by v_{hh}^n or v_{vh}^n using equation 1. Then v_s^n is dependant on γ which yields an equation where γ occures twice and can be iterated.

5.3   Results

Evaluating the shear wave velocities with above equations an estimated water content of 10% and an estimated specific unit weight of 26.5kN/m³ has been applied. This yields the following results:

Table 1: Evaluation of measurements before the installation of powder columns, Blasberg [3]

path	depth	polarisation	K_0	v_s^n	γ	e
	[m]		[]	[m/s]	[kN/m³]	[]
1	3	X	1.58	177.1	15.38	0.896
1	3	Z	1.58	176.1	15.31	0.904
2	3	X	1.23	179.6	15.56	0.873
2	3	Z	1.23	179.4	15.55	0.875
3	3	X	1.52	179.3	15.54	0.876
3	3	Z	1.52	178.5	15.48	0.883

Table 2: Evaluation of measurements after the installation of powder columns, Blasberg [3]

path	depth	polarisation	K_0	v_s^n	γ	e
	[m]		[]	[m/s]	[kN/m³]	[]
1	3	X	0.54	237.3	21.69	0.344
1	3	Z	0.54	235.3	21.40	0.362
2	3	X	0.89	222.3	19.68	0.481
2	3	Z	0.89	222.2	19.67	0.482
3	3	X	0.53	244.2	22.77	0.280
3	3	Z	0.53	242.1	22.44	0.299

Before the installation of powder columns the earth pressure coefficient is evaluated to be above 1 which is unexpected, Tab.1. Estimations yielded values around 0.5 because of the unloaded state. Void ratios of approximately 0.9 seem quite realistic. After the installation the earth pressure coefficient changed heavily. In the same time the void ratio decreased by factor two to three. It is noted, that the void ratios after the installation are smaller than the ones for the densest state for sand allone, because these are average values including the powder columns.

5.4   Discussion of results

The discussion of the results may be enhanced with the comparison of additional data from other field tests like borings and earth pressure measurements in test area A3.

Five borings have been executed and specimen have been taken which have been investigated in the laboratory concerning the type of soil, water content and density. The soil is mainly sandy and the average water contents lies between 12 to 16%. The densities show good agreement with the ones evaluated from the crosshole measurements.

Earth pressure measurements have been executed in test area A3 adjacent to test area A2 of the crosshole measurements. The average unit weight of area A3 is 15.8 kN/m^3 similar to the one in A2. An obvious difference exists in the fines content being approximately 30% higher than in area A2.

To investigate the earth pressure before, during and after the installation of one column, earth pressure spades have been installed around the column. Earth pressure coefficients between 0.6 and 2.2 have been measured before the installation of the first column showing a large deviation. The spades displayed totally different values being only some decimeters apart. The average value before the installation of the columns lies between 1.2 and 1.4 confirming the shear wave velocity measurements in area A2 where values between 1.2 and 1.6 have been measured.

After the installation of the first column all stresses are decreasing rapidly down to values between 0.2 and 0.7 being 0.5 as an average one. This behaviour seems to be similar to the one recorded during the shear wave measurements. In contrast to the crosshole measurements all values increase again to around 1.0 after the columns surrounding the first column have been installed.

The above measurements in test area A3 indicate, that the soil has been slightly constrained ($K_0¿1$) before the installation of the columns. This constraint has been lost during the vibration of the installation of the first column ($K_0=0.5$), despite a horizontal displacement of the soil around the column which seems not to have reached the earth pressure spades. Together with the installation of the other columns surrounding the first one, the soil between the columns has been densified in a way, that the pressures in the spades have increased again.

In test area A2 where the crosshole measurements have been executed, the earth pressure coefficients do not exibit high values after the column installation like in

A3 most likely due to the different fines content mentioned above. The fines content is related to the compressibility of a soil, determining the radius of compacted soil around a column. Because in both test areas the same weight of the vibrator string has been used, the same horizontal and vertical stresses have been applied during the column installation. The high percentage of fines in A3 leads to larger displacements without much volume change of the soil, and therefore the stresses have been transferred from the vibrator to the spades over a larger distance.

Finally it can not be proved be shear wave measurements alone, if the soil improvement is due to the installation of powder columns only or if the surrounding soil is improved as well. Additional RC-tests with fly ash would be necessary to recalculate the shear wave velocities in the fly ash and in the soil separately.

6 Summary

The application of crosshole measurements for compaction control in an inhomogeneous open pit mining area near Leipzip in Germany is described. The soil has been improved with powder columns by Keller Grundbau Gmbh.

Crosshole measurements before and after the installation of the columns have been used to prove the densification of the soil. Shear wave velocities in two different depths with three measurement paths each have been measured. Different shear wave directions have been used to determine the earth pressure coefficients and then the void ratio with the help of a stress normalisation.

Earth pressure coefficients around 1.4 have been evaluated before the installation of the columns and around 0.7 afterwards. The corresponding void ratios decreased from 0.9 to about 0.4, which shows the impressively the effect of the soil improvement.

A comparison with borehole logs and conventional earth pressure measurements yield, that the crosshole technique is a practically applicable method for compaction control.

Acknowlegdements

The help of Gerwin Blasberg for the compilation of the results and the execution of the field test is acknowledged. The same applied to Mike Schellhase and Joseph Benkler for the support of the field measurements. For the installation of all electronic devices I would like to thank Dr. Gerhard Huber und Rolf Steinbrunn and in the end the help of Dr. Peter Kudella is acknowlegded compiling the results of the conventional earth pressure measurements.

References

[1] BEHLE, A. (1972): *Modellseismische Untersuchung des Systems aus Rayleighwelle und Scher-welle an der Oberfläche des elastischen Halbraumes bei verschiedenen Herdtiefen Heft 16.* Hamburger Geophysikalische Einzelschriften, University Hamburg

[2] BELLOTTI, R.; JAMIOLKOWSKI, M. (1996): *Anisotropy of small strain Stiffness in Ticino sand.* Geotechnique 46, No.1, 115-131

[3] BLASBERG, G. (1998): *Bestimmung des Erddruckbeiwertes und der Porenzahl in einer bindigen Kippe mit dem Crosshole-Verfahren.* Unpublished thesis, Institute of soil mechnics and rock mechanics, University of Karlsruhe

[4] FIORAVANTE, V. ET AL. (1997): *Anisotropy of elastic stiffness in Kenya sand.* Geotechnique (submitted)

[5] GUDEHUS G. (1998): *On the onset of avalances in flooded loose sand.* Phil. Trans. Royal Society London A. Vol. 356, 2747-3761

[6] HAUPT, W. (1986): *Bodendynamik, Grundlagen und Anwendung.* Vieweg & Sohn Verlag Braunschweig/Wiesbaden

[7] HEELAN, P. A. (1953): *Radiation form a cylindrical source of finite length.* Geophysics 18, 685-696.

[8] HOAR, R.J.; STOKOE, K.H. (1978): *Generation and Measurement of Shear Waves In Situ.*

[9] INSTITUTE OF SOIL MECHNICS AND ROCK MECHANICS, UNIVERSITY OF KARLSRUHE (1996): *Sanierung und Sicherung setzungsfließgefährdeter Kippen und Kippenböschung* Annual report to LMBV

[10] JAMIOLKOWSKI, M.; LO PRESTI, D.C.F ET AL. (1994): *Validity of insitu tests related to real behaviour.* XIII ICSMFE, New Dehli, India, 51-55.

[11] JAMIOLKOWSKI, M.; LO PRESTI, D.C.F ET AL. (1997): *Assessment of coefficient of earth pressure at rest from shear wave velocity mearurements.* Geotechnique (in print)

[12] KLEIN, G. (1990): *Bodendynamik und Erdbeben.* Grundbautaschenbuch, Vierte Auflage, Ernst & Sohn Verlag, Berlin

[13] PRANGE, B. (1983): *Der Resonant-Column-Versuch; Theorie und Experiment.* Symposium: Meßtechnik in Erd- und Grundbau

[14] RAJU, V. (1994): *Spontane Verflüssigung lockerer granularer Körper - Phänomene, Ursachen, Vermeidung.* Institute of soil mechnics and rock mechanics, University of Karlsruhe

[15] ROESLER, S.K. (1979): *Anisotropic shear modulus due to stress anisotropy.* J. Geotech. Engng. Div. A.S.C.E. 105, GT5, 871-880

[16] SCHELLHASE, M. (1997): *Bestimmung der Porenzahl einer bindigen Kippe mit dem Crosshole-Verfahren.* Unpublished thesis, Institute of soil mechnics and rock mechanics, University of Karlsruhe

[17] STUDER, J.; ZIEGLER, A. (1986): *Bodendynamik.* Springer-Verlag Berlin Heidelberg New York Tokyo

[18] WEHR, W.; STEIN, U.; CUDMANI, R., BÖSINGER E. (1995): *CPT, shear wave propagation and freeze probing to estimate the void ratio in loose sands.* Int. Symposium on cone penetration testing CPT'95, Linköping, Vol. 2, 351-356

Compaction control with a dynamic cone penetrometer

Chaigneau L., Gourves R., Boissier D.

LERMES Université Blaise Pascal Clermont-Ferrand, France.

Abstract: The compaction control consists in measuring the dry density and compares it with the standard Proctor density. The cone resistance in a known granular medium is directly linked to the dry density. It is the comparison between the in-situ penetrogram and a reference curve which permits the compaction. This reference curve corresponds, for a given soil and a required compaction level to the cone resistance (for a passed compaction). A calibration process is necessary to establishes the reference curves.
This paper sets out the compaction control method and the calibration Process.

Introduction

The mechanical behaviour of a soil can be approached by two different ways. Either the granular medium is considered as a continuous medium with its constitutive laws, or the material is regarded as a discontinuous medium, through this micro mechanic approach, the global properties are deduced form the locale properties.

The grains parameter (mineralogy, grains, shape, distribution of contacts granulometric range, water content etc...) are used to deduce the macroscopic mechanical behaviour. The granular medium is considered as discontinuous and its grains are supposed unbreakable and insensitive to attrition. With these hypotheses, the granular assembly response to a given stress is dully defined. It is the case for the cone resistance if the dutch formula hypotheses are validated. In a granular medium for a saturation state, the cone resistance depends on dry density.

It is proposed to establish this relationship and to use this property in compaction control.

1 Penetrometric tests in an homogeneous medium

1.1   The penetrometer

The dynamic cone penetrometer with variable energy is used. The principle of the apparatus consists in driving with a hammer a shank string, which is 14mm in diameter with a 2cm² or 4cm² cone. (fig.1). The speed of impact and the depth are measured for each blow. The dynamic cone resistance q_d is calculated by using the Dutch formula. The cone resistance which q_d are function of depth, are automatically recorded [Gourves91] [Zhou97]. The results can be transferred to a computer to edit the penetrograms.

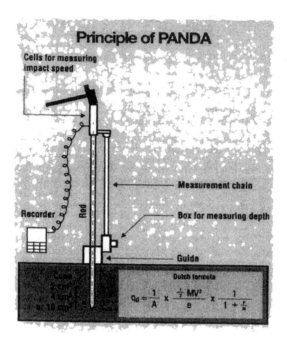

Figure 1: Principle of the PANDA

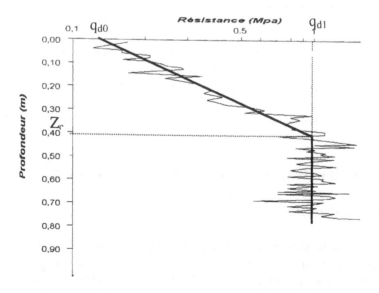

Figure 2 : example of a penetrogram in a sand with homogeneous density.

1.2 Classical penetrogram

Many studies have established [Gourves et Al 1995] [Gourves et Zhou 1997] that in a granular medium with is homogeneous density and water content, the penetrogram can be modelled using a system of co-ordinates (log q_d; z) by 2 lines defined by three parameters (fig.2): q_{d0} the cone resistance at the surface, Z_c the critical depth and q_{d1} the cone resistance below the critical depth.

2   Compaction control with a dynamic cone penetrometer

Many studies on dynamic cone penetrometer have proved the extreme sensibility of cone resistance to the dry density. These apparatus are not only well adapted for the detection lack of compaction, but they are also suited to detect the number and the thickness of the in-situ compacted layers. The compaction control with a penetrometer for a given material and a water content is based on the establishment of the relationship between cone resistance q_d and the dry density. With an accurate soil classification enough, for each material contained in a same soil class, the mechanical behaviour is quite the same (the french soil classification GTR [SETRA94] is used in our example). A database set up, for each type of soil, a relationship between the dry density and the parameter of penetrograms (q_{d0}, Z_c, q_{d1}) in an homogeneous medium for different waters contents.

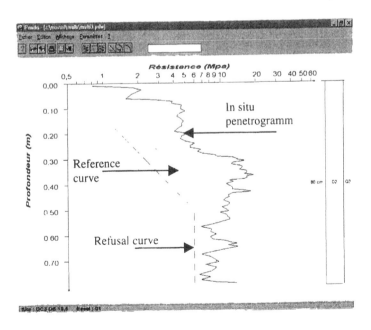

Figure 3: reference and refusal penetrogram.

For a given material and a water content, the cone resistance for the required compaction level is determinated by the previous relationship. The compaction control is achieved by comparing the in-situ penetrogram to a reference curve. Another penetrogram is added to the reference one; this penetrogram is called refusal curve. The refusal curve corresponds about 98% of the reference density. The refusal curve takes into account variations in properties within the same granulometric class, the errors in the relationship between dry density and q_d, the experimental errors, and an allowance to fail or not the compaction (fig.3).

3 Calibration Process

3.1 Principle

The principle of the PANDA calibration Process consists in establishing a database. This base contains identified materials according an accurate soil classification. This classification must contain three main type of parameter to characterise the soil behaviour: the granulometric range of the material (in particular the fraction passing the 80µm and the 2mm sieve, d_{min}, D_{max}, the coefficient of curvature, and the coefficient of Hazen), the average shape of grains and for the fine material the water sensitivity (fraction passing the 400µ sieve). The water sensitivity is measured by the Atterberg limits or by the blue methylene value. This database includes the relationship between dry density and the cone resistance for different materials and water contents.

3.2 Mould specification

The PANDA cone is very small (2cm²) and requires a limited volume of soil. The calibration process could be done in a rigid mould. This one must have sufficient dimensions to avoid border effects which increase the cone resistance. Then mould must have 35cm in diameter (more the 20 tip diameter) [Cassan] [Robertson] and a sufficient height to observe the critical depth. The calibration moult is 37.4 cm in diameter and 80.6cm height.

3.3 Homogeneous media achievement

Once the material characterised, it is introduced into a mould to compact it at known water content and dry density by static compaction (odometric way). A mould compacted by this method gives a good homogeneity, which is better than those obtained by dynamic compaction layer by layer. [De Carvalho87]. The main difficulties of this method concerns the lateral friction between the soil and the surface of the mould. The friction increases with the pressure, then the stress field is not constant inside the soil. A density gradient appears in the height of the mould. To thwart this effect, the compaction in achieved by two layers and the surface of the mould is lubricated [Chaigneau98] (the soil is protected by a thin plastic membrane). With this method, the frictions are sufficiently limited to obtain an homogeneous density mould. The static compaction of this mould needs a 1000kN hydraulic press (fig.4).

Figure 4: Hydraulic Press.

3.4   Reference curve establishment

If the soil is water sensitive, different tests are achieved with different water contents. The chosen water intervals are : [0.7 w_{sp} and 0.9 w_{sp}]; [0.9 w_{sp} and 1.1 w_{sp}] and [1.1 w_{sp} and 1.3 w_{sp}] (w_{sp}: water content at the standard Proctor density). For each material and each moisture content, the soil is compressed to five levels of density. The first density corresponds to the bulk density and the last one the maximum density which is possible to obtain with the press. This maximum obtained is about 110% of the standard Proctor density. This density interval covers the densities, which are most commonly used in backfill compaction. The relationships between cone resistance and dry density are drawn (γ_d-q_{d1}). This curves are modelled by a logarithmic regression ($\gamma_d = a \ln(q_{d1})+b$) (fig.5).

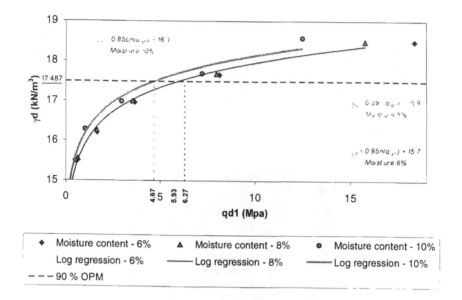

Figure 5: relationship between dry density and cone resistance q_{d1} below the critical depth.

The cone resistance qd1 is defined for each compaction level (95%, 98% 100% of the proctor density). A tolerance value is calculed (2% of tolerance on the dry density correspond to 40% on qd1). The cone resistance at the surface (q_{d0}) and the critical depth (Zc) are determinated by the same way.

4 Conclusion

The compaction method proposed is very reliable and was not only used in fine soil but also in coarse soil (up to 50mm). The database contains the non-evolutive materials, which are most commonly used in backfill in the world.

It is possible to control compaction on a large depth (several meters) with a great accuracy to detect the efficiency of a compaction in depth, this is the main interest of the method.

Actually, the database based on the french GTR classification is going to change, to be used with the main international classifications.

References

Cassan, M. (1988) « Les essais in situ en mécanique des sols ». Tome 1 réalisation et interprétation. Eyrolles, pp103-125.

Camapum De Carvalho, J., Crispel, JJ., Mieuddens, C. and Nardone, A.(1987) « la reconstitution des éprouvettes en laboratoire ». Rapport de recherche LPC N°145.

Chaigneau, L. (1998) « Protocole de calibration du pénétromètre analogique numérique dynamique assisté par ordinateur : PANDA. Mise au point d'un mode opératoire de laboratoire ». Rapport de D.E.A de l'université Blaise Pascal, France.

Gourves, R. (1991) « Le PANDA – pénétromètre dynamique léger à énergie variable pour la reconnaissance des sols ». document interne Université Blaise Pascal, Clermont-Ferrand, France.

Gourves, R. et Zhou, S. (1997) «The in situ characterization of the mechanical properties of granular media with the help of penetrometer». 3rd International conférence on Micromécanique of granular média, Duram pp 57 - 60.

Gourves, R. et Richard, B. (1995) «Le pénétromètre dynamique léger PANDA». proceeding of ECSMFE. Copenhagen, Denmark pp 83 - 88.

LCPC. SETRA (1994). « Remblayage des tranchées et réfection des chaussées ». Guide technique.

Robertson, P.K. (1992) « cone penetration testing in geotechnical pratice ». E&FN Spon. pp 291-304.

Zhou. S. (1997) « caractérisation des sols de surfaces à l'aide du pénétromètre dynamique léger à énergie variable type 'PANDA' ». Thèse de doctorat de l'université Blaise Pascal, Clermont-Ferrand, France.

3 On-line compaction control

Sophisticated roller compaction technologies and roller-integrated compaction control

Dietmar Adam and Fritz Kopf

Institute for Ground Engineering and Soil Mechanics, Technical University of Vienna, Austria

Abstract: In this paper compaction technologies are considered in form of rollers equipped with different kinds of exciters. Compaction of soil, fill material, etc. normally takes place by vibratory rollers, whereby the vibration of the drum is caused by a rotating mass excenter. In the last years dynamic rollers with different types of excitation where developed. The paper is focused on the oscillatory roller, the VARIO roller, and the automatically controlled VARIO CONTROL roller. Furthermore, the paper contains roller-integrated compaction control methods in order to check continuously the compaction success during compaction with different dynamic roller types. Finally, the application of new compaction technologies and case histories, where roller-integrated continuous compaction controls (CCC) were used, are reported.

1 Introduction

The quality of roads, highways, motorways, rail tracks, airfields, earth dams, waste disposal facilities, foundations of structures and buildings, etc. depends highly on the degree of compaction of filled layers consisting of different kinds of materials, e.g. soil, granular material, artificial powders and grain mixtures, unbound and bound material. Thus, both compaction method and compaction equipment have to be selected taking into consideration the used material suitable for the prevailing purpose. Compaction process should be optimised in order to achieve sufficient compaction and uniform bearing and settlement conditions. If compaction control can be included in the compaction process, time can be saved and cost reduced. Furthermore, a high-levelled quality management requires continuous control all over the compacted area, which can only be achieved economically by roller-integrated methods.

2 Roller compaction

Rollers (Fig. 1) are widely used for compacting soil, layers of filled material, etc. Depending on
- layer thickness to be compacted,

and material properties like
- grain size distribution, maximum grain size and grain shape,
- water content,
- water and air permeability

rollers are selected, whereby following machine parameters mainly contribute:

- Total roller weight and static drum load
- Static compaction (without vibration) or
- Dynamic compaction caused by different kinds of excitation with parameters:
  - ➤ Direction of resulting dynamic contact force
  - ➤ Excitation frequency
  - ➤ Theoretical drum amplitude
- Surface shape and diameter of drum

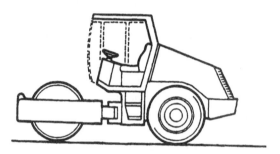

Figure 1: Self-propelled single drum roller

2.1 Static rollers

Rollers with static drums use the effective dead weight of the machine to apply pressure to a particular surface. Thus, soil particles are pressed together and the void content is reduced. Adequate compaction of static rollers is normally achieved only in the upper layers of the material, i.e. the depth effect of static compaction is limited.

Cohesive fine grained soil can be compacted sufficiently with static rollers in connection with padfoot drums (see also 2.3.2). By means of the low permeability of fine grained soil pore water pressures are created by applying (dynamic) compressive stresses. Pore pressures reduce the compaction effect significantly or prevent compaction at all. However, a statically passing padfoot drum "remolds" the soil near the surface resulting in a reduction of pore pressures and void ratio respectively. Nevertheless, only thin layers can be compacted sufficiently.

2.2 Dynamic rollers

Dynamic rollers make use of a vibrating or oscillating mechanism, which consists of one or more rotating eccentric weights. During dynamic compaction a combination of dynamic and static loads is used. The dynamically excited drum delivers a rapid succession of impacts to the underlying surface from where the compressive and shear waves are transmitted through the material to set the particles in motion. This eliminates periodically the internal friction and facilitates in combination with the static load the rearrangement of the particles into positions that result in a low void ratio and a high density respectively. Furthermore, the increase in the number of contact points and planes between the grains leads to higher stability, stiffness, and lower long-term settlement behaviour.

2.2.1 Vibratory roller

The drum of a vibratory roller is excited by a rotating excenter mass which is shafted on the drum axis (Fig. 2). The rotating mass sets the drum in a circular translatoric motion, i.e. the direction of the resulting force is corresponding with the eccenter position. Compaction is achieved mainly by transmitted compression waves in combination with the effective static drum load. Consequently, the maximum resulting compaction force is supposed to be almost vertical and in fact it is only a little inclined.

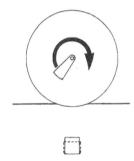

Figure 2: Excitation of a vibratory roller drum and dynamic compaction effect (compression)

The vibration of the roller drum changes in dependence of the soil response. Numerous investigations have revealed that the drum of a vibratory roller operates in different conditions depending on roller and soil parameters. Five operating conditions specified in table 1 can occur; definition criteria are on the one hand the contact condition between drum and soil and on the other hand the drum cycle as a multiple of the excitation cycle [1].

Table 1: Operating condition of a vibratory roller drum

Behaviour of motion	Interaction drum-soil	Cycle *)	Operating condition	Application of CCC	Soil stiffness
periodic	contact	1	CONTINUOUS CONTACT	yes	low
	partial loss of contact	1	PARTIAL UPLIFT	yes	
		2 (4)	DOUBLE JUMP	yes	
		[2(4)]	ROCKING MOTION	no	
chaotic		-	CHAOTIC MOTION	no	high

*) Specified as a multiple of excitation cycle $T=2\pi/\omega_0$.

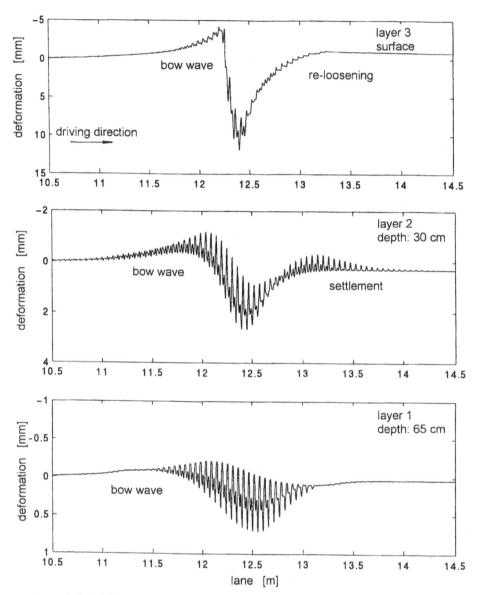

Figure 3: Soil deformation in three depths according to the impact of a vibratory roller drum
operating in the condition double jump

Continuous contact only occurs when the soil stiffness is very low, i.e. in case of uncompacted or soft clayey layers. Partial uplift and double jump (Fig. 3) are the most frequent operating conditions. The difference between these two operating conditions consists of the number of excitation cycles; consequently, the motion behaviour of the drum repeats itself. When the soil stiffness increases further the motion of the drum axis is no more vertical, the drum starts rocking. Very high soil stiffness in combination with disadventageous roller parameters can cause chaotic motion of the drum. While rocking and chaotic drum motion the roller is not manoeuvrable any more. No useful compaction is possible then. Vibratory rollers are the mainly used rollers world-wide. They can be employed universally for a wide range of soil types and granular material.

2.2.2 Oscillatory roller

The drum of an oscillatory roller oscillates torsionally. The torsional motion is caused by two opposite rotating excenter masses, which shafts are arranged excentrically to the axis of the drum. Thus, soil is loaded horizontally in addition to the vertical static dead load of the drum and the contributing roller frame. These cyclic and dynamic horizontal forces result in additional soil shear deformation; dynamic compaction is achieved mainly by transmitted shear waves (Fig. 4).

Figure 4: Excitation of an oscillatory roller drum and dynamic compaction effect (shearing)

Investigations have revealed that oscillatory rollers operate in two conditions depending on roller and soil parameters. If the force exceeds the friction force (incl. the adhesion) between drum and soil the drum starts slipping. While slipping the compaction effect is reduced, however, the surface is "sealed" by the slip motion [14]. Consequently, oscillatory rollers are mainly employed for asphalt compaction and cohesive material. Furthermore, oscillatory rollers are used in the vicinity of sensitive structures, because the emitted vibrations are significantly lower than those of vibratory rollers.

2.2.3 VARIO roller

The VARIO roller is a development of the BOMAG company. In a VARIO roller two counter-rotating exciter masses, which are concentrically shafted on the axis of the drum, cause a directed vibration. The direction of excitation can be adjusted by turning the complete exciter unit in order to optimise the compaction effect for the respective soil type (Fig. 5). If the exciter direction is (almost) vertical or inclined, the compaction effect of a VARIO roller can be compared with that of a vibratory roller. However, if the exciter direction is horizontal, the VARIO roller compacts soil like an oscillatory roller, although the motion behaviour of the drum is different. The shear deformation of soil is caused by a horizontally translatoric motion, whereas the drum of an oscillatory roller is working torsionally. Thus, a VARIO roller can be used both for dynamic compression compaction (like a vibratory roller), for dynamic shear compaction (like an oscillatory roller), and a combination of these two possibilities, depending only on the adjustable force direction (Fig. 6). Consequently, VARIO rollers can be employed universally for each soil type, the respective optimum direction can be found by basic investigations on a test field on site [12].

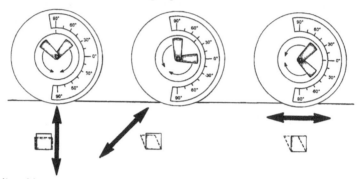

Figure 5: Adjustable excitation direction of a VARIO roller drum and respective compaction effect

2.2.4   VARIO CONTROL roller

Based on the findings relating to the ways of operating of different dynamic rollers, the BOMAG company developed the first automatically controlled so called VARIO CONTROL roller [9, 16]. In this roller type the direction of excitation (vibrations can be directed infinitely from the vertical to the horizontal direction) is controlled automatically by using defined control criteria, which allow an optimised compaction process and, consequently, a highly uniform compaction (Fig. 7). The control criteria are explained in the following [15]:

*   Operating criterion
    If the drum passes to the operating condition double jump, the excitation direction is immediately changed, so that the operating condition partial uplift is kept.

*   Force criterion
    If the specified maximum compaction force is reached, the excitation direction is changed by the automatic control system, so that the applied force does not exceed the maximum force (Fig. 8).

Two accelerometers, which are mounted on the bearing of the drum, record the dynamic motion behaviour continuously. Soil contact force, energy delivered to soil, displacements, etc. are calculated in a processor unit taking into consideration the roller parameters like masses, exciter force and frequency, etc. The data are transmitted to an integrated control system, where the setting of the parameters is managed automatically.

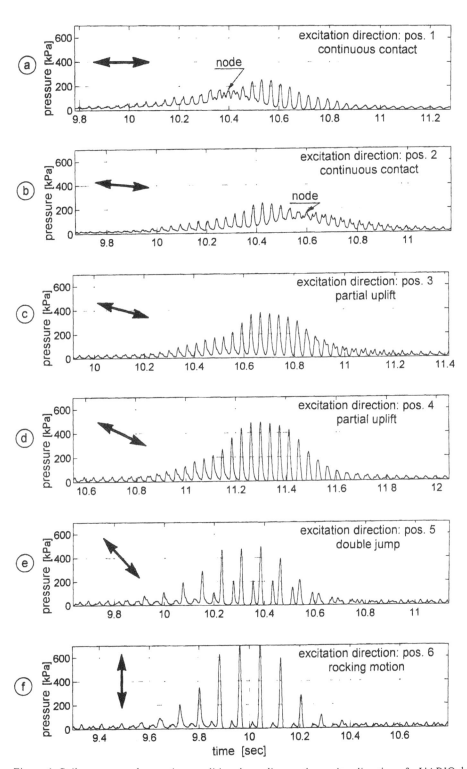

Figure 6: Soil pressure and operating condition depending on the exciter direction of a VARIO drum

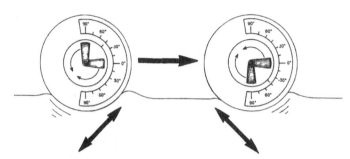

Figure 7: Direction of inclination for optimum compaction taking into account the driving direction

Following significant benefits of VARIO CONTROL rollers have been revealed by practical application:

- Uniform compaction by continuous adjustment of force direction
- Optimised compaction combined with less roller passes
- Prevention of over-compaction and re-loosenings
- Improved compaction both in deeper layers and on surface
- Reduction of lateral vibrations in the vicinity of sensitive structures

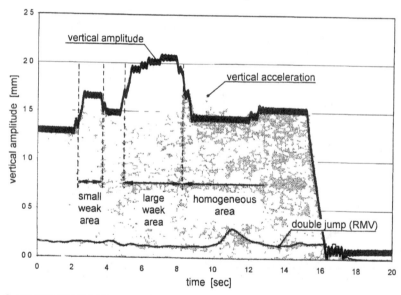

Figure 8: VARIO CONTROL automatic control of excitation direction depending on soil properties

BOMAG has developed the so called VARIOMATIC roller for automatically controlled asphalt compaction already in 1997 [9].

The development of automatically controlled rollers is a challenge for roller producers [10]. The Swiss company AMMANN has already constructed an auto-controlled roller in connection with a roller-integrated control system in order to determine dynamic compaction values independent from roller parameters [7].

2.3 Drum types with various surface shape

In order to optimise the compaction process the surface shape of the drum has to be considered among the selection of the roller and the kind of drum excitation. In the following some basic aspects of various drums are shown and fields of applications are specified.

2.3.1   Smooth drum

Coarse grained and medium grained soils are usually compacted with smooth drums. By means of the cylindrical shape of the drum the contact area changes with the drum penetration and, consequently, the applied pressure. A further effect of the smooth curved drum is the occurrence of a "bow wave" in front of the drum and a "back wave" behind (Fig. 3).

2.3.2   Padfoot drum

Cohesive fine grained soil usually cannot be compacted sufficiently with smooth drums. By means of the low permeability of fine grained soil pore water pressures are created by applying (dynamic) compressive stresses. Pore pressures reduce the compaction effect significantly or prevent compaction at all. However, a statically passing padfoot drum "remolds" the soil near the surface resulting in a reduction of pore pressures and void ratio respectively (Fig. 9). Furthermore, "bow waves" and "back waves" are prevented by the surface shape of the padfoot drum.

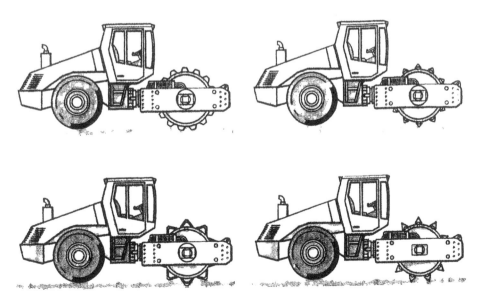

Figure 9: Padfoot drum (left above) and three different kinds of specially shaped drum surfaces developed for various purposes [13]

2.3.3   Drums with special shape

In the last years specially shaped drum surfaces were developed in dependence of the used material. Thus, hardly compactable soils, soft rocks, slates, and sandstone could be compacted efficiently with dynamic and static rollers equipped with specially shaped drum surfaces (Fig. 9).

A basic research project concerning the optimisation of a specially shaped drum ("polygon drum") was recently started. Project partners are the BOMAG company and the Technical University of Vienna, Institute for Ground Engineering and Soil Mechanics.

3   Continuous compaction control (CCC)

Hitherto, compaction control has been carried out mainly by means of punctual test methods with the purpose to check the density or stiffness of the compacted layer. Common testing methods, like the sand equivalent, the water balloon method or the *TROXLER*-nuclear gauge test to determine the density and the load plate test for determining the soil stiffness, are punctual methods. The uniformity cannot be approved, and faults can hardly be detected by such spot tests. Moreover, the measuring depth range is about 20 to 40 cm only. Last but not least all spot test methods are relatively expensive. This testing also delays construction work due to the fact that construction activities must not be carried out in the vicinity of a spot test because ground vibrations could affect the test results.

Therefore, these conventional methods of compaction control are not sufficient any more for high quality projects. The roller-integrated continuous compaction control (CCC) represents an improvement and is based on the measurement of the dynamic interaction between dynamic rollers and soil [1].

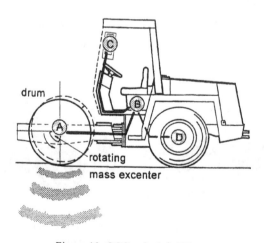

Figure 10: CCC-principle [3]

3.1 CCC-principle

As mentioned in chapter 2.1 the motion behaviour of different dynamically excited roller drums changes in dependence of the soil response. This fact is used to determine the stiffness of the ground. Accordingly, the drum of the dynamic roller is used as a measuring tool; its motion behaviour is recorded (Fig. 10: A), analysed in a processor unit (Fig. 10: B) where a dynamic compaction value is calculated, and visualised on a dial or on a display unit (Fig. 10: C) where data can also be stored. Furthermore, an auxiliary sensor is necessary to determine the location of the roller (Fig. 10: D). Control data are already available during the compaction process and all over the compacted area.

3.2 CCC-systems

Two recording systems are available for vibratory rollers and VARIO rollers with vertical or any inclined excitation direction (except horizontal direction). One recording system has been developed for oscillatory rollers.

All systems consist of a sensor containing one or two accelerometers attached to the bearing of the roller drum, a processor unit and a display to visualise the measured values. The sensor continuously records the acceleration of the drum. The time history of the acceleration signal is analysed in the processor unit in order to determine dynamic compaction values with regard to specified roller parameters.

3.2.1   Compactometer

Compactometer is a product of the Swedish company Geodynamik AB; the first patent was taken out in 1978.

The Compactometer Value (CMV) is calculated by dividing the amplitude of the first harmonic of the acceleration signal by the amplitude of the exciting frequency (1):

$$CMV \sim \frac{\hat{a}(2\omega_0)}{\hat{a}(\omega_0)} \qquad\qquad RMV \sim \frac{\hat{a}(0{,}5\omega_0)}{\hat{a}(\omega_0)} \qquad (1, 2)$$

Former empirical investigations have revealed that the amount of the first harmonic increases with increasing soil stiffness. Moreover, an auxiliary value was created in order to take into consideration the operating conditions of the drum. The Resonant Meter Value (RMV) is calculated by dividing the amplitude of the half frequency of the acceleration signal by the amplitude of the exciting frequency (2). The operating conditions influence the progress of CMV significantly (Fig. 11).

3.2.2   Terrameter

Terrameter is a product of the German BOMAG company. CCC-technique based on the Terrameter calculates the energy absorbed by the soil. The acceleration of the drum $\ddot{z}_1$ is measured in two orthogonal directions, the drum velocity $\dot{z}_1$ is determined by integrating the acceleration components. Taking into consideration the mass of drum m_D, the rotating mass excenter m_E, the static force $\vec{F}_{stat}$ and the excenter force $\vec{F}_E$ the energy W_{eff} is calculated as follows (3):

$$OMEGA \sim W_{eff} = \oint_{2T} \left[-(m_D + m_E)\ddot{z}_1 + \vec{F}_{stat} + \vec{F}_E \right] \dot{z}_1 \, dt \qquad (3)$$

The integration is performed by two cycles of excitation to take the operating condition of double jump into account (Fig. 11).

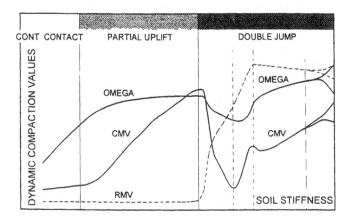

Figure 11: Progress of CMV and OMEGA depending on soil stiffness and operating conditions [2]

3.2.3   Oscillometer

Oscillometer is the only existing CCC-system for oscillatory rollers and is a product of the Swedish company Geodynamik AB [11].

The Oscillometer Value (OMV) results from the horizontal acceleration signal of the (sinusoidally) oscillating drum $\ddot{x}$ (Fig. 12, 13). By means of the partially slipping behaviour of the drum, the response is different from a sinusoidal signal. Investigations have revealed that the gradient of the response signal at zero is corresponding with soil stiffness. Thus, OMV is calculated from the first derivation of the acceleration signal at $\ddot{x} = 0$ of each cycle with period t_{per}:

$$OMV = \left| \frac{d\ddot{x}}{dt} t_{per} \right| \qquad at : \ddot{x} = 0 \qquad (4)$$

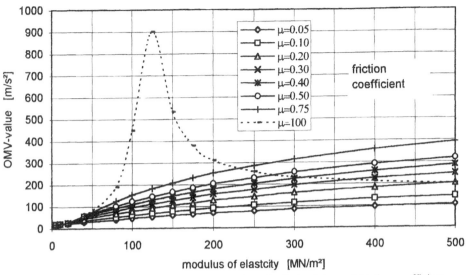

Figure 12: Progress of OMV depending on modulus of soil elasticity and friction coefficient

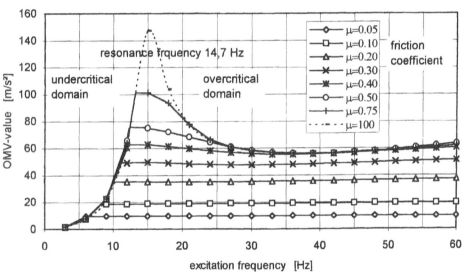

Figure 13: Progress of OMV depending on excitation frequency and friction coefficient

3.3 CCC-criteria

The dynamic compaction values are relative values having a clear physical background. If the data shall be compared with common conventional values like the deformation modulus of the static load plate test or the soil density calibrations have to be performed. There are several possibilities to select the spots where conventional tests can be carried out. Best correlation can be achieved if the spots are selected by means of CCC-results. Spots with high, intermediate, and low dynamic compaction values indicate a wide range of soil properties. Nevertheless, different operating conditions have to be taken into account else completely irrelevant regression lines are determined (Fig. 14).

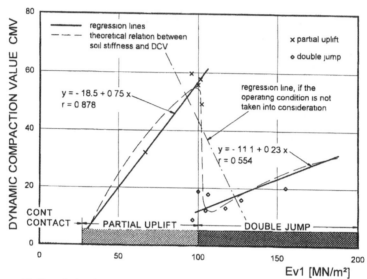

Figure 14: Correlation between E_{v1} and CMV taking into account operating conditions

Furthermore, the different depth range of CCC and the conventional test methods must be considered. In general the measuring depth of CCC (depending on the total roller weight) is larger than the compaction depth and the measuring depth of common spot tests. Thus, soft soils in deeper layers can be detected with CCC which is not possible with conventional tests. In Fig. 15 a typical CCC-result is shown. The information provided is used to set up control criteria:

- Minimum value in order to locate weak spots and areas
- Maximum values in order to locate areas with highest soil stiffness
- Mean value in order to assess the general condition of the checked area
- Standard deviation in order to assess the uniformity of the checked area
- Increase of the dynamic compaction value in order to point out the further compactability
- Decrease of the dynamic compaction values as indication for loosening and/or grain crushing

The use of CCC for acceptance testing requires contractual limits for minimum, maximum and mean values. These can be set up by carrying out correlation tests. Extensive investigations and practical site experience have revealed that the standard deviation shall not be > 20% (related to the mean value), the increase of dynamic compaction values between two passes < 5%.

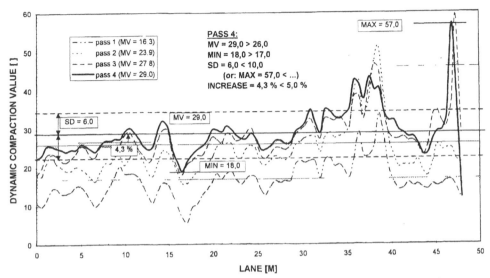

Figure 15: Progress of CCC-data and CCC-criteria: MV mean value, MIN minimum value,
MAX maximum value, SD standard deviation, INCREASE between two roller passes

3.4 Austrian CCC-standard

In Austria a new standard for road and motorway construction concerning CCC
was established in 1998 and declared obligatory in 1999. This RVS 8S.02.6
("Kontinuierlicher walzenintegrierter Verdichtungsnachweis") [17] contains both
fields of application, requirements to roller, soil, CCC-systems, and contractual
limits for acceptance testing including the setting up of these limits by calibration,
distribution of responsibility and cost.

3.5 Fields of CCC-application

In the last years CCC was applied successfully within the scope of quality
management and control for many projects in Austria, Germany, Sweden and
other European countries. Furthermore, the new method was introduced to
countries all over the world. In the beginning CCC was limited to sub-bases and
base courses in road construction, later CCC-application was extended to a wide
range in different fields:

- Motorways and highways
- Rail tracks
- Airfields
- Earth and rock fill dams
- Waste disposal facilities
- Foundations of structures and buildings

4   Case histories in Austria

4.1 Deponie Asten (Upper Austria)

A new concept for the deposit of squeezed sewage sludge required the extension of the existing disposal. The base liner was to be made up by a fine-grained hydrogenating material (treated fly ash *REALIT*). The thickness of the layer amounted to 25 cm in the base and 30 cm at the slopes, which had an inclination of 67 % [4, 5].

The compaction process was optimised by application of CCC to test fields both in the plain and at the slope. After having documented the subsoil with the roller integrated method the treated fly ash (mixed in plant) was placed and compacted. Consequently, the compaction was controlled by CCC and the compaction process was fully and continuously documented. Due to the high inclination of the slope the roller was pulled by an excavator, which was situated on the crown of the slope (see Photo 1).

Photo 1: Asten. Vibratory roller pulled by an excavator. CCC is performed during compaction

Criteria for sufficient compaction were constituted by

- mean values of each lane,
- the progress of the dynamic compaction values and
- the increase of measured data between two roller passes.

4.2 Deponie Mitte Unterfrauenhaid (Burgenland)

The waste disposal facility in Unterfrauenhaid was extended between 1998 and 1999 (Fig. 16). It is situated in the area of a former sand pit. The exploited pit was refilled with excavation material but not compacted. Soil exploration has revealed that the artificial underground was not suited as base for a waste disposal facility. Thus, the refilled material was excavated and filled in 30 cm thick layers again. The demand for a work integrated and continuous control required the application of CCC. Furthermore, CCC was used for acceptance testing of mineral base liners. Conventional control methods like permeability testing and use of *TROXLER*-nuclear gauge equipment could be minimised. Thus, time could be saved and cost reduced significantly.

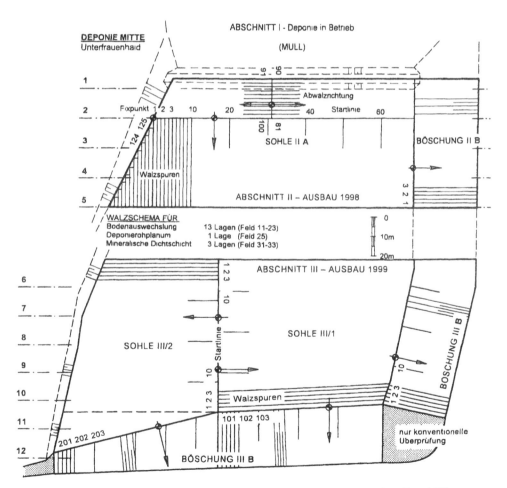

Figure 16: Situation of waste disposal facility Unterfrauenhaid (extension sections II and III). Organisation of roller lanes for compaction and roller-integrated CCC

4.3 Schutzdämme Schwaz-Eiblschrofen (Tyrol)

The rock slip disaster on the Eiblschrofen near Schwaz in July 1999 brought a village in danger of rock impacts. Thus, it was decided to construct rock-fill dams in order to protect the village. The dams had to be situated in areas of major danger, so that construction works could be carried out only having taken special precautions.

A further necessity was to construct the dams as fast as possible, nevertheless, the maximum quality was demanded. On this conditions the quality management was based on roller-integrated continuous compaction control with the Terrameter. The calibration of CCC-value OMEGA and the layer thickness were set up on a test field outside the danger zone. The quality control was performed work-integrated while compacting.

Criteria for sufficient compaction were constituted by

- minimum values for small and large amplitude (indicator for material quality and water content) and
- the increase of measured data between two roller passes (indicator for sufficient compaction).

If the measured value exceeded the required minimum value and, furthermore, the increase of OMEGA was less than 5% related to the mean value of the last pass on this lane compaction work could be finished on this lane.

4.4 Kraftwerk Donaustadt (Vienna)

In the power plant Donaustadt in Vienna a new power station block is under construction. The extension is situated in an area with unsuitable ground properties. The over-consolidated tertiary soil in a depth of about 12 m to 14 m below surface is overlain by loose quaternary gravel. The upper layers with a total thickness of about 3 m to 4 m consist of loose sands and fillings. By means of high loads, irregular load distribution, and the existence of dynamically loaded machine foundations highly uniform soil properties have been required.

Thus, the foundation concept has been based on the following steps:

- Removing of sands and artificial fillings 4 m below surface;
- Application of vibroflotation soil compaction to quaternary gravel layer from excavation level to tertiary soil;
- Soil exchange from excavation level up to base of slab. Sandy gravels from Vienna subway excavation pits serve as fill material.

Quality control of fill material is based on CCC with Terrameter. In Fig. 17 the process of OMEGA achieved on the excavation level after having applied vibroflotation method is shown. Deep compaction points can be seen significantly in the process of OMEGA. The CCC-result will become more and more uniform with increasing material thickness.

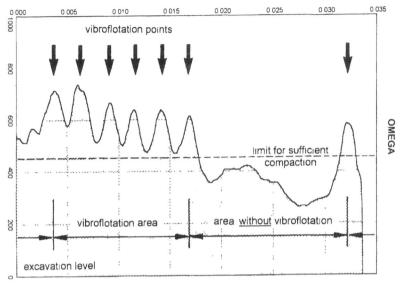

Figure 17: Power plant Donaustadt. Progress of OMEGA achieved on excavation level

5 Conclusions

Sophisticated roller compaction technologies provide a wide range of possibilities to select the adequate roller for the respective purpose. The classic vibratory roller operating in five different conditions is supplemented by rollers with different kinds of excitation. The torsionally behaving oscillatory roller is specially suitable for asphalt compaction, cohesive mineral liners and in the vicinity of sensitive structures. In a VARIO roller the vibrating excitation is directed; the direction can be adjusted depending on the soil properties, so that optimised compaction can be achieved. Thus, VARIO rollers can be employed universally for each soil type and purpose. A further development is the automatically controlled roller VARIO CONTROL, whereby the direction of excitation is controlled automatically by using defined control criteria. VARIO CONTROL compaction provides uniform compaction, less roller passes, improved compaction both in deeper layers and on surface, and reduction of lateral vibrations, e.g. for employment in the vicinity of sensitive structures. Among the roller type the surface shape of the drum should be considered in order to optimise compaction effect. Smooth drums are usually employed for compacting coarse grained and medium grained soil. Cohesive fine grained soil is better compacted by using padfoot drums. Specially shaped drums were developed for respective purposes. Roller-integrated continuous compaction control (CCC) represents an improvement for high-levelled quality management systems. Compaction control is integrated in the compaction process and data are provided all over the compacted area. Two measurement systems are available for vibratory and VARIO rollers (Compactometer and Terrameter), one recording system has been developed for oscillatory rollers (Oscillometer).

References

[1]   ADAM, D. (1996): *Flächendeckende Dynamische Verdichtungskontrolle (FDVK) mit Vibrationswalzen*. Dissertation. Institut für Grundbau und Bodenmechanik, TU Wien.

[2]   ADAM, D., BRANDL, H. (1997): *Roller-Integrated Continuous Compaction Control of Soils*. Proc. of 3rd International Conference on Soil Dynamics (ICSD-III), August 3-7, 1997, Tiberias, Israel.

[3]   ADAM, D. (1997): *Continuous compaction control with vibratory rollers*. Proc. of GeoEnvironment, 1st Australia – New Zealand Conference on Environmental Geotechnics, 245 – 250, November 26-28, 1997, Melbourne, Victoria, Australia.

[4]   ADAM, D., KOPF, F., MACHO, T. (1998): *Anwendung der Flächendeckenden Dynamischen Verdichtungskontrolle (FDVK) auf steilen Deponieböschungen*. Mitteilung des Institutes für Grundbau und Bodenmechanik, Heft 56, Technische Universität Braunschweig.

[5]   ADAM, D., KOPF, F. (1998): *Application of continuous compaction control (CCC) to waste disposal liners*. Proc. of Third International Congress on Environmental Geotechnics, September 7-11, 1998, Lisboa, Portugal.

[6]   ADAM, D., KOPF, F. (1998): *Messungen der Verformung in Böden zufolge statischer und dynamischer Einwirkungen*. Mitteilung des IGB, Heft 55, TU Braunschweig.

[7]   ANDEREGG, R. (1997): *Nichtlineare Schwingungen bei dynamischen Bodenverdichtern*. Dissertation, Diss. ETH Nr. 12419, Eidgenössische Technische Hochschule Zürich.

[8]   BRANDL, H., ADAM, D. (1997): *Sophisticated Continuous Compaction Control of Soils and Granular Materials*. Proc. of the XIVth International Conference on Soil Mechanics & Foundation Engineering, Vol. 1: 31-36, September 6-12, 1997, Hamburg, Germany.

[9]   FLOSS, R., HENNING, J. (1998): *VARIOMATIC - Ein entscheidender Schritt zur Qualitätssicherung im modernen Erd- und Verkehrswegebau*. BOMAG BA 049, Deutschland.

[10]   GEODYNAMIK AB (1995): *Control of a compacting machine with a measurement of the characteristics of the ground material*. Patentschrift WO 95/10664.

[11]   GEODYNAMIK AB (1997): *Verfahren zum Bestimmen des Verdichtungsgrades beim Verdichten einer Verdichtungsmaschine u. Vorrichtung zur Durchführung des Verfahrens*. (Oszillometer). Patentschrift DE 35 90 610.

[12]   HENNING, J., UEBERBACH, K.-O. (1996): *Moderne Verdichtung*. Straße und Tiefbau 11/96.

[13]   KLOUBERT, H.-J. (1999): *Anwendungsorientierte Forschung und Entwicklung löst Verdichtungsprobleme im Erd- und Asphaltbau*. Tiefbau 12/1999, Deutschland.

[14]   KOPF, F. (1999): *Flächendeckende Dynamische Verdichtungskontrolle (FDVK) bei der Verdichtung von Böden durch dynamische Walzen mit unterschiedlichen Anregungsarten*. Dissertation. Institut für Grundbau und Bodenmechanik, TU Wien.

[15]   KOPF, F., ADAM, D. (1999): *Optimierte Verdichtung mit selbstregelnden Walzen*. Tagungsband, 2. Österreichische Geotechniktagung, 22.-23. Februar 1999, Wien.

[16]   KRÖBER, W. (1999): *VARIO CONTROL und FDVK im Erdbau – schwierige Verdichtungsaufgaben sicher und wirtschaftlich gelöst*. Tagungsband: "Wachstum durch Innovation", 24-26. Februar 1999, Braunlage, Deutschland.

[17]   Richtlinien und Vorschriften für den Straßenbau RVS 8S.02.6 (1999): *Erdarbeiten. Kontinuierlicher walzenintegrierter Verdichtungsnachweis*. FSV im ÖIAV, Wien.

Quality control in deep vibrocompaction

Wolfgang Fellin

Institute of Geotechnics and Tunnelling, University of Innsbruck, Austria

Abstract: Deep vibrocompaction (vibroflotation) is a method of ground improvement up to depths of 40 m. It has been successfully used since 1936.

The major problem of this method is, that, inspite of its good performance, nobody knows exactly how the soil gets compacted. The methods of compaction control during the vibration are unreliably. The achieved compaction can only be verified after vibration.

In this paper it is shown that information from the movement of the vibrator can be used as additional quality control and indicator for the degree of compaction. Such information can be analysed directly during compaction and offers thereby the possibility of an "on-line compaction control".

1 Introduction into deep vibrocompaction

Sand and gravel with less than 25% fines and less than 5% clay can be compacted with deep vibrocompaction.

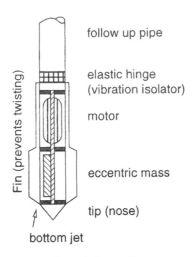

Figure 1: Deep vibrator

The vibrator is a cylindrical tube that houses eccentrically rotating masses. These rotating masses induce a horizontal vibratory motion (see figure 1). The tube diameter is between 25 and 42 cm, its length between 2.7 and 4.4 m, the weight between 800

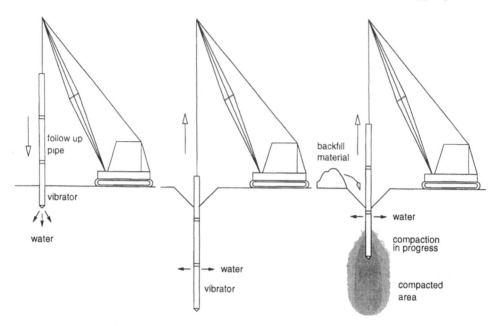

Figure 2: Procedure of deep vibration compaction

and 2600 kg. The induced centrifugal force is 150 up to 472 kN with a frequency from 25 up to 60 Hz.

The vibrator is lowered due to its vibration and its own weight (see figure 2). Water or air jetting can increase the penetration rate. Typical penetration rates are 1.0 up to 2.0 m/min. When the required depth is reached, the water flow is redirected to the upper jets. There are two methods for withdrawing the vibrator:

- Vibrating in a constant depth for a fixed time (30 to 60 s) or until a fixed amount of power consumption is reached. Then the vibrator is withdrawn between 0.3 and 1 m to the next depth.

- Withdrawing the vibrator by 0.3 to 1 m, then lowering again by half the withdrawn depth or until a fixed amount of power consumption is reached or until the vibrator is not able to sink anymore (pilgrim step method).

Backfill material is added into the developing crater. Up to 1.5 m^3/m depth might be required. The usual maximal depth is about 25 m but sometimes it is possible to reach 40 m. The first 3 m are poorly compacted. They have to be removed or compacted using other compaction methods, e.g. heavy tamping or roller compaction.

The compacted soil column of a single vibration point has a radius of 1.5 to 3 m. To compact an area the vibrator is lowered at points with a particular pattern as sketched

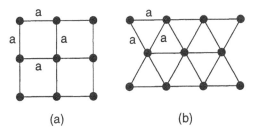

Figure 3: Pattern of compaction points: (a) rectangular pattern for spread footings, (b) triangular pattern for large areas

in figure 3. The spacing a between the compaction points is usually between 1.5 and 3.0 m. Patterns for spread footing have lower spacings than patterns for large areas. The finer the sand is, the smaller is the spacing.

For further information see e.g. [3, 11, 2, 6, 1, 5]

2 Compaction control - state of the art

Verifying the compaction success is quite difficult because the density of cohesionless soil cannot be measured directly. Other quantities like cone penetration resistance have to be used to determine the achieved density.

Compaction control is split into two parts:

Monitoring during vibration: During compaction the following parameters are recorded over the depth:

- Quality and amount of backfill material
- Maximum and time history of power consumption
- Time needed for each compaction point

The power consumption is usually used to deduce the achieved density. This is by no means sufficient!

Compaction control after vibration: The achieved density is tested by cone penetration or standard penetration tests in some centers of the triangles or rectangles of the compaction pattern. It should be noted that the penetration resistance measured immediately after compaction can be up to 50% smaller than the penetration resistance measured one month later. This is the case mainly in fine sands below ground water level.

Measurements of the amplitude of the tip of the vibrator using the the pilgrim step method in minor depth (5 m) showed, that during withdrawal the amplitude correlates with the penetration resistance of tests after compaction [10]. Actually no company uses this additional information.

3   Mechanical model of vibration

Vibrocompaction is influenced by two major components. The vibrator on one hand and the surrounding soil on the other hand. If the surrounding soil changes its properties during vibration the motion of the vibrator should be changed. The main idea in developing an on-line compaction control is to study the motion, measure this motion and deduce from the measurements the properties of the soil.

3.1   Vibrator suspended in air

The simplest model of a vibrator oscillating in air is the physical pendulum sketched in figure 4. The plane motion of this pendulum is a projection of the rotation of the vibrator.

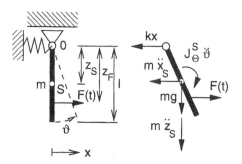

Figure 4: Vibrator in air

The equation of motion of the system in figure 4 is:

$$
\begin{aligned}
m\ddot{x} + mz_S\ddot{\vartheta} + kx &= F\cos\Omega t \\
mz_S\ddot{x} + (J_\Theta^S + mz_S^2)\ddot{\vartheta} + gmz_S\vartheta &= z_F F\cos\Omega t
\end{aligned}
$$

Herein m is the mass of the vibrator, k is the stiffness of the elastic hinge, z_F the distance of the center of gravity and J_Θ^S the moment of inertia around the center of gravity. Amplitudes calculated with $k = 0$ are a good approximation of the real ones given by the companies. The frequency Ω of the excitation is much higher than the natural frequency ω of the system.

With these assumptions ($k = 0$, $\Omega \gg \omega$) the amplitude of the tip of the vibrator oscillating in air is

$$A_L \approx \frac{F}{\Omega^2}\left(\frac{1}{m} + \frac{(z_F - z_S)(l - z_S)}{J_\Theta^S}\right) .$$

With the additional approximation that the mass is uniformly distributed and the excitation force is $F = m_u r \Omega^2$ (a rotating mass m_u with the eccentricity r) the amplitude is approximately

$$A_L \approx 2.5 \frac{m_u r}{m}$$

It is important to note, that not only the tip has an amplitude. Also the shoulder (below the hinge) deflects.

3.2 Vibrator embedded in soil

Like in classical soil dynamics the surrounding soil is modelled by springs and dash-pots. Two models are possible:

- Analogy to foundation oscillation

- Analogy to pile oscillation

3.2.1 Analogy to foundation oscillation

A vibrator embedded in soil can be modelled in analogy to foundation oscillation (see figure 5).

The equations of motion are:

$$m\ddot{x}_S + c\dot{x} + kx = F\cos\Omega t \tag{1}$$
$$J_\Theta^S \ddot{\vartheta} + c_\vartheta \dot{\vartheta} + (k_\vartheta + mgz_S)\vartheta = (z_F - z_S)F\cos\Omega t \tag{2}$$

The motion of the center of gravity is

$$x_S = X_S\sin(\Omega t + \alpha_S) , \quad X_S = \frac{F}{\sqrt{(k - m\Omega^2)^2 + c^2\Omega^2}}$$

$$\alpha_S = \arctan\left(\frac{k - m\Omega^2}{c\Omega}\right) \quad \text{with} \quad -\frac{\pi}{2} \leq \alpha_S \leq \frac{\pi}{2} .$$

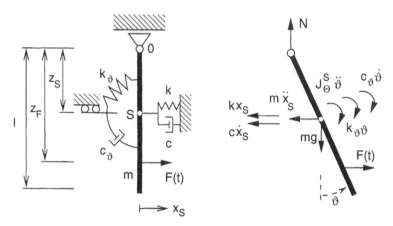

Figure 5: Analogy to foundation oscillation (schematic)

The rotation is

$$\vartheta = \theta \sin(\Omega t + \alpha_\vartheta) \,, \; \theta = \frac{(z_F - z_S)F}{\sqrt{(k_\vartheta + mgz_S - J_\Theta^S \Omega^2)^2 + c_\vartheta^2 \Omega^2}}$$

$$\alpha_\vartheta = \arctan\left(\frac{k_\vartheta + mgz_S - J_\Theta^S \Omega^2}{c_\vartheta \Omega}\right) \quad \text{with} \quad -\frac{\pi}{2} \le \alpha_\vartheta \le \frac{\pi}{2} \,.$$

Since $\alpha_S \approx \alpha_\vartheta$ for usual vibrators the motion of the tip is

$$x = A\sin(\Omega t + \alpha) \,, \; A \approx X_S + (l - z_S)\theta \,, \; \alpha \approx \frac{\alpha_S + \alpha_\vartheta}{2} \tag{3}$$

3.2.2 Analogy to pile oscillation

With the assumption that the stiffness of the vibrator is much higher than the stiffness of the soil the vibrator can be modelled like in figure 6. The soil is modelled with a dynamic Winkler interaction, with continuously distributed springs k_x and dashpots c_x [9].

The equations of motion are:

$$m\ddot{x}_S + c_x l\dot{x} + c_x le\dot{\vartheta} + k_x lx_S + k_x le\vartheta = F\cos\Omega t$$

$$J_\Theta^S \ddot{\vartheta} + c_x le\dot{x}_S + c_x l\left(\frac{l^2}{12} + e^2\right)\dot{\vartheta} + k_x lex_S + \left(k_x l\left(\frac{l^2}{12} + e^2\right) + mgz_S\right)\vartheta$$
$$= (z_F - z_S)F\cos\Omega t$$

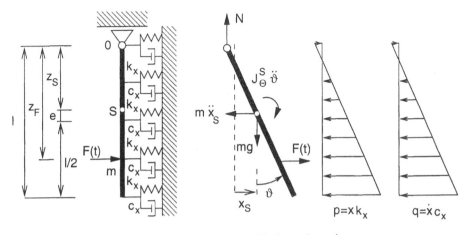

Figure 6: Analogy to pile oscillation (schematic)

For usual vibrators the center of gravity is almost in the middle of the vibrator, so $e \approx 0$. In that case the equations reduce to the equations of the model with foundation analogy (1) and (2) with:

$$k = lk_k \;,\; c = lc_x \;,\; k_\vartheta = k_x \frac{l^3}{12} \;,\; c_\vartheta = c_x \frac{l^3}{12}$$

3.2.3 Approximation for usual vibrators

With the assumption that the mass of the vibrator is uniformly distributed, $J_\Theta^S = m\frac{l^2}{12}$, and neglecting the moment of the weight $mgz_S \ll lk_x \frac{l^2}{12} = k_\vartheta$ the motion of the vibrator is:

$$x_S = X_S \sin(\Omega t + \alpha) \;,\; X_S = \frac{F}{\sqrt{(k - m\Omega^2)^2 + c^2\Omega^2}}$$

$$\vartheta = \theta \sin(\Omega t + \alpha) \;,\; \theta = 6\frac{2z_f - l}{l^2} X_S$$

$$\alpha = \arctan\left(\frac{k - m\Omega^2}{c\Omega}\right)$$

The amplitude of the tip is

$$A = \left(6\frac{z_F}{l} - 2\right) \frac{F}{\sqrt{(k - m\Omega^2)^2 + c^2\Omega^2}} \;.$$

3.2.4 Model parameters, springs and dashpots

The spring and dashpot constants can be calculated first for small strains. Then the influence of large strains can be iteratively taken into account.

Small strains: First the shear wave velocity has to be calculated [4]:

$$c_s = (19.70 - 9.06e)\sigma_0'^{\frac{1}{4}} \quad \text{if} \quad \sigma_0' \geq 95.8 \text{ kN/m}^2$$

$$c_s = (11.36 - 5.35e)\sigma_0'^{\frac{1}{4}} \quad \text{if} \quad \sigma_0' < 95.8 \text{ kN/m}^2$$

Here e is the void ratio and $\sigma_0' = (\sigma_1' + \sigma_2' + \sigma_3')/3$ the effective mean stress.

The shear modulus of the soil is $G = c_s^2\rho$. The density of the soil ρ can be calculated by $\rho = \rho_s/(e + 1)$, with the grain density ρ_s of the sand.

Further the compression wave velocity is needed

$$c_p = \frac{c_s}{\sqrt{\frac{1-2\nu}{2-2\nu}}} \;,\quad \nu = \frac{K_0}{(1 + K_0)} \;,\quad K_0 \approx 1 - \sin\varphi \;.$$

The Poisson ratio ν should be calculated with the coefficient of earth pressure at rest K_0. K_0 can be approximated by the use of the friction angle of the soil φ.

The spring and dashpot constants for the motion of a vibrator with the length l and the diameter D are (horizontal motion [12, p. 33, Tabelle 2-2 2-3A], rotation [12, p. 69]):

$$k = \frac{G\frac{D}{2}}{1 - \nu}\left[3.1\left(\frac{l}{D}\right)^{0.75} + 1.6\right] \;,\quad c = \rho c_p D l$$

$$k_\vartheta = K_\vartheta^{stat}\left(1 - \frac{1}{3}\frac{b_0^2}{1 + b_0^2}\right) \;,\quad c_\vartheta = \frac{K_\vartheta^{stat}}{3}\frac{b_0^2}{1 + b_0^2}$$

Therein $b_0 = z_0\Omega/v_p$, $z_0 = 9/32\,r_0\pi(1-\nu)\,(v_p/v_s)^2$ and $r_0 = \sqrt[4]{4I_S/\pi}$. The static stiffness is [7]

$$K_\vartheta^{stat} = \frac{GI_S^{0.75}}{1 - \nu}\left(\frac{l}{D}\right)^{0.25}\left[2.4 + 0.5\frac{D}{l}\right] \;.$$

The area moment $I_S = Dl^3/12 + Dl\,(z_S - l/2)^2$.

The constants for a horizontal oscillating pile given in [9] lead for this problem to unrealistic small amplitudes. In the pile analogy it is better to use

$$k_x = \frac{k}{l} \;,\quad c_x = \frac{c}{l} \;. \tag{4}$$

Large strains: When large strains are taken into account the shear modulus has to be reduced and the material damping has to be included [12]:

$$G = \beta c_s^2 \rho \; , c = c_p Dl + 2D_m \frac{k}{\Omega} \; , \; c_\theta = \frac{K_\vartheta^{stat}}{3} \frac{b_0^2}{1 + b_0^2} + \frac{l^2}{12} 2D_m \frac{k}{\Omega}$$

Therein β can be found as function of the soil and the shear strains in [8, p. 242, figures 7-10]. E.g. $\beta = 0.3$ for sand and $\gamma = 10^{-3}$. The damping D_m is [4, p 290, figure 8.19] for small strains ($\gamma = 10^{-6}$) approximately $D_m = 0.05$, and increases for large strains ($\gamma = 10^{-2}$) to $D_m = 0.26$. The shear strains can be estimated as $\gamma \approx A\Omega/v_p$, where A is the amplitude of the tip [5, p 52].

3.3 Example

The vibrator V42 of the company VIBROFLOTATION

mass m	length l	diameter D	frequency	$F = m_u r \Omega^2$	J_Θ^S	z_S	z_F
2592 kg	2.99 m	38.4 cm	25 Hz	472 kN	1662 m^{2}kg	1.57 m	2.19 m

is vibrating in a depth of 10 m in sand with $\gamma_s = 26$ kN/m^3 and the coefficient of earth pressure at rest is $K_0 = 0.5$. The shear strains are estimated to 10^{-3}, so the damping is $D_m = 0.16$ and the reduction of the shear modulus is $\beta = 0.3$.

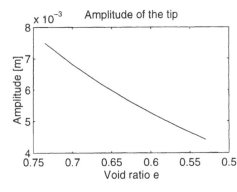
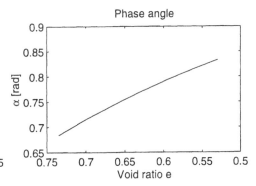

Figure 7: Amplitude of a V42 vibrator operating in sand

Figure 8: Phase angle of a V42 vibrator operating in sand

The results of equations 3 in figures 7 and 8 show a decreasing amplitude for increasing density (decreasing void ratio). Also the phase angle is changing significantly within the range of void ratio. The eccentric mass is leading the motion of the vibrator by the phase angle $\varphi = \pi/2 - \alpha$ [5, p. 57]. Measurements of the amplitude and the phase angle of the leading eccentric mass should be an indicative for the achieved density.

4   On-line compaction control

On-line compaction control[1] can be provided by measuring the horizontal accelerations in two orthogonal directions at the tip and at the shoulder. Herefrom the amplitudes can be obtained by integration. Further the phase angle of the eccentric mass has to be determined. This can be done by a pulse given by an emitter when the eccentric mass passes a certain point of the vibrator tube. From the time lag of this pulse to the zero points of the time history of the accelerations the phase angle can be calculated.

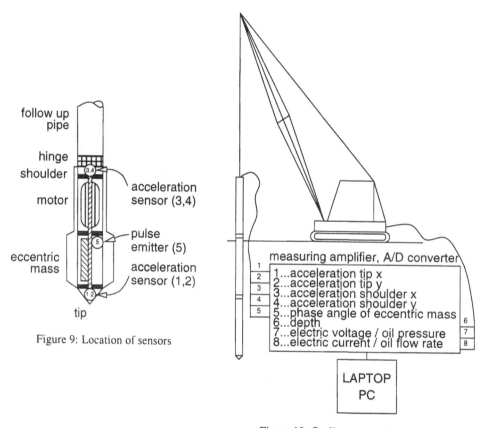

Figure 9: Location of sensors

Figure 10: On-line compaction control

With the signals of the five sensors placed as shown in figure 9 it is possible to describe the motion of the vibrator completely. The change of amplitudes and phase angle in addition to the measured electric current or oil pressure can be used to determine the change of density during vibration.

---

[1]The on-line compaction control is applied for patent number 19928629.2.

5   Quality control

It is very important for the success of the deep vibrocompaction that the vibrator has a permanent contact with the soil. The material down flow around the vibrator must be high enough to replace the displaced and compacted soil (so called "self feeding"). Is the material down flow too low, the vibrator starts to strike out a cavity and its centrifugal force is not well transmitted into the soil anymore. This drastically reduces the compaction effect.

If the vibrator is continuously in contact with the soil, its movement is more or the less circular (or elliptic, due to the fins). If the vibrator loses the permanent contact it impacts and rebounds off the wall of the cavity and the motion becomes irregular.

For circular or elliptic motion a frequency analysis of the acceleration signals shows only the first harmonic in the frequency spectrum. If the motion becomes irregular other harmonics appear. This can be used to detect the loss of continuous contact. A higher water supply should then help to increase the material down flow.

Altogether the on-line compaction control and the frequency analysis can be used for a quality control of deep vibrocompaction. The additional measurements give much more information about the compaction success during vibration as the presently used current or oil pressure measurement.

Summary

It is strongly recommended to measure more than current or oil pressure during vibration. Firstly because the verification of the compaction success after vibration is too time consuming, especially when the cone resistance increases with time over weeks. Secondly because the penetration test are done only in a few points. This gives no continuous information over the whole compacted area. And third, if the penetration test shows bad results, it is a lot of effort to bring the whole equipment back to the poorly compacted area, and sometimes the vibrator is even not able to penetrate again.

The deep vibrocompaction equipment should have a "green signal lamp" which indicates when a certain soil density is achieved during deep vibration compaction. It is shown that this should be possible by measuring the amplitudes of the vibrator at its tip and shoulder, as well as the phase angle of the leading eccentric mass (see figure 7). Conclusions about the compaction success can be drawn by analysing these values during vibration.

Irregular motion, which is most likely due to a loss of permanent contact between soil and vibrator, can be detected by a frequency analysis of the acceleration signals.

Thus, the measurements suggested in this paper can be used (additionally to the values measured so far) for the *on-line compaction control* and quality control.

References

[1] B. Broms. Deep compaction of granular soils. In Hsai-Yang Fang, editor, *Foundation Engineering Handbook*, chapter 23, pages 814–832. Chapman & Hall, 2 edition, 1991.

[2] R. Brown. Vibroflotation compaction of cohesionless soils. *Journal of the Geotechnical Engineering Division, ASCE*, 103:1437–1451, 1977.

[3] E. D'Appolonia. Loose sands - their compaction by vibroflotation. *American Society for Testing Materials, Special Technical Publication No. 156*, pages 138–154, 1953.

[4] B.M. Das. *Fundamentals of Soil Dynamics*. Elsevier Science Publishing, 1983. ISBN 0-444-00705-9.

[5] W. Fellin. *Rütteldruckverdichtung als plastodynamisches Problem*, volume 2 of *Advances in Geotechnical Engineering and Tunnelling*. A.A. Balkema, 2000.

[6] FG Straßenwesen. Merkblatt für Untergrundverbesserung durch Tiefenrüttler. Forschungsgesellschaft Straßenwesen, Köln: Straßenbau AZ. Sammlung technischer Regelwerke und amtlicher Bestimmungen für das Straßenwesen. Stand Oktober 1984, Abschnitt Untergrundverbesserung, 1979.

[7] G. Gazetas. Foundation vibration. In Hsai-Yang Fang, editor, *Foundation Engineering Handbook*, chapter 15, pages 553–593. Chapman & Hall, 2 edition, 1991.

[8] W. Haupt. Dynamische Bodeneigenschaften und ihre Ermittlung. In W. Haupt, editor, *Bodendynamik, Grundlagen und Anwendung*, chapter 7, pages 225–279. Friedrich Vieweg & Sohn. ISBN 3-528-08878-8, 1986.

[9] N. Makris and G. Gazetas. Displacement phase differences in a harmonically oscillating pile. *Géotechnique*, 43(1):135–150, 1993.

[10] J.G.D. Morgan and G.H. Thomson. Instrumentation methods for control of ground density in deep vibrocompaction. In *Proceedings of the 8th European Conference on Soil Mechanics and Foundation Engineering*, Helsinki, 1983.

[11] M. Poteur. *Beitrag zur Untersuchung des Verhaltens von Böden unter dem Einfluß von Tauchrüttlern*. PhD thesis, Fakultät für Bauwesen der technischen Hochschule München, 1968.

[12] J.P. Wolf. *Foundation Vibration Analysis Using Simple Physical Models*. PTR Prentice-Hall, ISBN 0-13-010711-5, 1994.

4 Quasi static experimental investigations, cyclic loading

Material fabric and mechanical properties after compaction for unbounded granular materials

Gunther Gidel[1,2] , Denys Breysse[1] , Alain Denis[1] , Jean-Jacques Chauvin[2]

[1] Centre de Développement des Géosciences Appliquées, Université Bordeaux I, France
[2] Laboratoire Régional des Ponts et Chaussées de Bordeaux, France

Abstract: When unbounded granular materials prepared using vibrocompression are subjected to many cycles of loading at various q/p ratios, it appears a wide scatter due to the variability of the material fabric. To improve the prediction of permanent strains, this scatter must be reduced. That's why an experimental program has been focused on the material fabric observed after compaction. The statistical treatment of the results shows that the initial compactness (e. γ_d) is firstly influenced by the aggregates type and the intensity of compaction when the initial stiffness (E_1) and the permanent stains (ϵ^p) mainly depend on the water content. The same approach will be developed for a large amount of cycles to predict the in-service response of unbounded granular materials.

1   Introduction

One of the main objectives of the road pavement design methods is to limit the development of ruts in the pavement structure. French method of design is based on too simplistic criteria, in particular with respect to unbounded granular materials used in the basic layer (for example the criterion of rutting verification depends only on traffic conditions)[1]. In fact, it doesn't take into account the real response of the different material components and too often results in unsatisfactory design (with overestimated or underestimated section). The "Laboratoires des Ponts et Chaussées (LPCs)", with the theme "Dimensionnement des chaussées souples" (Flexible pavement design), develop studies on the pavement materials to optimize their design. Our study focuses on the permanent strains of unbounded materials under triaxial cyclic loading (TCR tests)[2]. The long-term behavior has been poorly investigated essentially because of the too long duration of tests (4 days for each specimen, and usually 80 0000 cycles of load per specimen under each stress level) and the important experimental scattering (tests are performed on cylindrical specimens, 16 cm in diameter, 32 cm in height, of reconstructed and humidified granular materials, rebuilt by vibrocompression). In order to simplify the long-term behavior study, we have developed a new test-procedure with several successive stress levels [3]. For stress paths with q/p constant, and with moderate stress levels, it has been assumed that the effect of the previous cycles at lower stress levels are equivalent to the effect of only some cycles at the current stress level, if they induce the same permanent strain. This test, with successive stress levels, reduces the number of tests, and it provides a high level of information since it gives the permanent deformation for several stress levels from only one specimen. The LRPC in Toulouse and the LCPC in Nantes have also tested this test-procedure firstly developed in the LRPC in Bordeaux. They obtained good results, and built a relation between the axial permanent strain and both number of load cycles and stresses. Nevertheless, this relation assumes that all specimens are initially similar, which is not true. In fact, vibrocompression is an incompletely controlled fabrication process. During this operation, aggregates tend to closer and provide the specimen its initial stiffness and compactness. A double series of tests, which has been defined in the framework of experimental design, has been planned. The first series enables us to define the influence of the level of compaction and of the material characteristics (like the moisture content, the fillers content, the kind of aggregates) [4] on the initial state of the specimen: fabric and mechanical properties. The second series will help us in understanding how permanent strains develop under different stress levels, the initial state (stiffness / compactness) of the material being given. This document first presents the test program and then gives some results about the initial state parameters.

2   Definition of the first series of tests

This first series of tests is a stage of characterization of the initial state of the material before loading (TCR test with a large number of cycles). The determination of the initial state of the specimens is essential for a good estimation of the future evolution of the permanent strains. The results of the five different limestone materials classification (with respect to the French norm NF P 98-125), have shown that the stiffness and the permanent strains at the first cycles of loading strongly depend on this initial state. The shape of the curves is similar for all the materials, but the values of the stiffness and the permanent strains vary between the materials and between the specimens for a given material. Thus a series of tests must enable us to define the parameters of the initial state by the material characteristics (nature and constitution) and the level of compaction. This paragraph first presents the test-program chosen and next presents the tests included in this test-program.

2.1   Definition of the test-program

2.1.1   Fabric parameters

To accurately characterize the initial state of the material, we must take into account the material characteristics and the material fabric. The purpose is to define the surface response of the initial state parameters from the material characteristics and the material fabric. This study concerns only limestone materials coming from quarries in Charente-Maritime (France). Two materials are chosen for this test-program (five materials were initially available and we have privileged for this study the one, which had the best behavior and the one, which had the worst behavior, in accordance with the classifying method with TCR test). The physical meaning of the material parameter (note M) is unclear since, when only two materials are studied (A and B), it is not possible to say that M is the dry density, or the porosity, or the mechanical strength. Thus, depending on the stage of the study, physical reasons may be invoked to change the meaning of the M parameter which will appear in the relations. To be honest, this method is relevant only as long as the different properties describing each aggregate are highly correlated (in simple words, a given aggregate is more or less good). Several tests of characterization performed on the five aggregates types (porosity, density, mechanical strength) have shown they were strongly correlated ($R^2 > 86$ %). The influence of the fillers contents has been studied as well (parameter noted F). For each material (A and B) two different fillers content are taken, 10 % and 15 % (for each F-value the two materials have the same grading curve). All the specimens are built by vibrocompression, with an initial water content defined by the modified Proctor test. The fabric density usually is 97 % of the optimum modified Proctor density. In order to study the influence of the level of compaction a second fabric density equal to 100 % of the optimum modified Proctor density has been taken (parameter noted % OPM). Then the specimens are dried until the water content reaches target values defined by $W-W_{OPM}$ (parameter noted W). The unbounded granular materials behavior strongly depends on the water content, that's why four values have been taken for W. Table 1 summarizes the different values for each parameter. Normalized values in the interval [-1;+1] are also given. They will be used in the regression models.

Table 1: Definition of the different values for each parameter

Parameter	1	2	3	4
W	-1,5	-2	-2,5	-3
(normalized value)	(+1)	(+0,33)	(-0,33)	(-1)
F	10 %	15 %		
(normalized value)	(-1)	(+1)		
M	A	B		
(normalized value)	(-1)	(+1)		
% OPM	97 %	100 %		
(normalized value)	(-1)	(+1)		

2.1.2 Test-program

A study with all these parameters usually needs to perform 32 tests (4x2x2x2). Using the method developed by Taguchi [5] this number can be reduced to 16 tests. The complete test-program is presented on Table 2. It enables us to develop models using the four parameters and could also account for three interactions between these parameters. Each specimen is prepared according to the fabric conditions defined by the test-program. The initial state will be quantified regarding compactness and stiffness. A gammadensimetry test is performed to know the density and compactness of each specimen after compaction. Then a TCR test is performed to determine the stiffness induced by the compaction and measure the permanent strains for a low amount of load cycles.

Table 2: Complete test-program defined in according to the the Taguchi method

N	W	F	M	% OPM
1	1	1	1	1
2	1	1	2	2
3	1	2	1	2
4	1	2	2	1
5	2	1	1	2
6	2	1	2	1
7	2	2	1	1
8	2	2	2	2
9	3	1	1	2
10	3	1	2	1
11	3	2	1	1
12	3	2	2	2
13	4	1	1	1
14	4	1	2	2
15	4	2	1	2
16	4	2	2	1

3 Results and modelling

3.1 Gammadensimetry

This test is used to determine the density of a specimen. A radioactive source emits a gamma particle beam (C_o is the number of the emitted gamma particles). Some of them are absorbed by the specimen (μ' is the absorption coefficient of the material)

and a receiver measures the number of particles which have gone through the specimen (noted C). The density of the specimen is calculated by (1) (d is the thickness of the specimen).

$$\gamma = \frac{1}{\mu' d} \ln(\frac{C_o}{C}) \, .$$ (1)

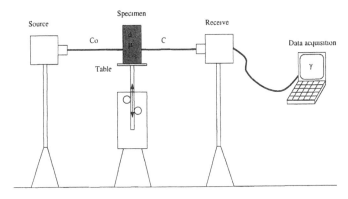

Figure 1: Apparatus for the density measurement of the specimen by gammadensimetry

Figure 1 represents the apparatus for the density measurement of the specimen by gammadensimetry.

The dry density is calculated by (2) where w is the water content of the specimen.

$$\gamma_d = \frac{\gamma}{1 + w} \, .$$ (2)

The compactness of the specimen is calculated by (3) where γ_s is the density of the aggregates (specific weight).

$$e = \frac{\gamma_s - \gamma_d}{\gamma_d} \, .$$ (3)

A specific attention has been given to the repeatability of the measurements. The vertical profiles of γ are given on Figure 2 for two specimens.

They clearly show that the density and the compactness of the specimens strongly depend on the material and the compaction intensity. It can also be noted that a certain variability exists within each specimen. This point has been analyzed but it is not developed in this paper. The surface response is derived from the multiple regression analysis on parameters and interactions. Only the factors which, after a Student test, are significant at a 5 % level, are kept in the regression models given here. For each model, the factors are ranked from the more influent to the less influent.

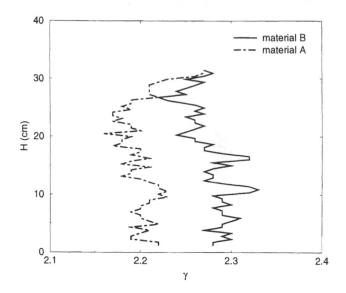

Figure 2: Vertical profiles for two specimens

The dry density can be modeled by (4).

$$\gamma_d = 2,071 + 0,044M + 0,017\%OPM + 0,017F .\qquad(4)$$

γ_d is greater for the material B than for the material A because the aggregates B are denser than the aggregates A ($\gamma_s = 2.38$ for B and 2.24 for A, here M means specific weight). The larger the fillers content, the higher the dry density of the specimens. The density also increases when the intensity compaction increases (more material is introduced for the same final volume).

The compactness can be modeled by (5).

$$e = 0,114 + 0,010M - 0,009\%OPM - 0,008F + 0,004\%OPM - M .\quad(5)$$

Like the dry density, the compactness depends on the material and the intensity of compaction, when compared to (4), the effect of the aggregates weight has been eliminated using (3). Therefore, the physical meaning of M is more probably linked with the angularity of the aggregates. The materials with angular aggregates are uneasy to compact and this induces worth compactness than materials with rounded aggregates. Figures 3 and 4 compare the experimental values and the values given by the models for the dry density and the compactness (squares represent a test out of test-program repeated four times to analyze the ability of the model to predict something).

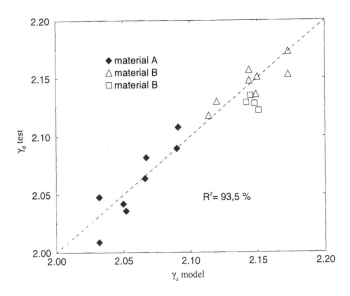

Figure 3: Comparison between the experimental and theorical values for γ_d

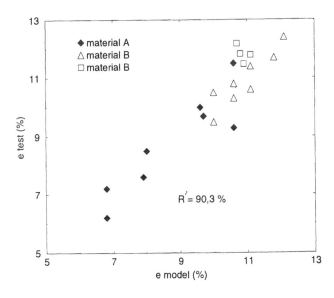

Figure 4: Comparison between the experimental and theorical values for e

These tests have shown the influence of the nature of the material (density of the aggregates, angularity, and fillers content) and the intensity of the compaction on the dry density and the compactness of the specimens. The parameter e eliminates the effect of the density of the aggregates, which makes it possible to distinguish the influence of the density of the aggregates from the influence of their angularity on the dry density.

3.2   TCR test

The TCR test consists in applying a cyclic loading on a specimen of unbounded material and in measuring its strains (resilient and permanent). The stiffness of the specimen is calculated by (6) under the assumption that the material is orthotropic and with $\nu_{vh} = 0,4$.

$$E_1 = \frac{\sigma_1 - 2\nu_{vh}\sigma_3}{\epsilon_1^r} \ .$$

(6)

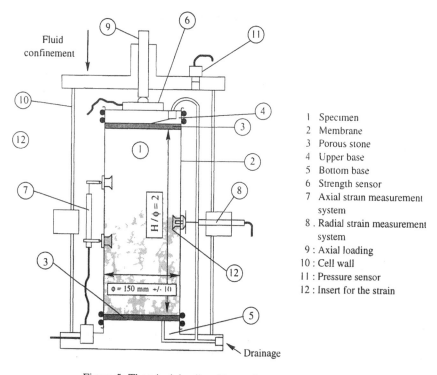

1   Specimen
2   Membrane
3   Porous stone
4   Upper base
5   Bottom base
6   Strength sensor
7   Axial strain measurement
    system
8 . Radial strain measurement
    system
9 : Axial loading
10 : Cell wall
11 : Pressure sensor
12 : Insert for the strain

Figure 5: The triaxial cell and its strain measurement system

Figure 5 represents the triaxial cell and its strain measurement system.

The results clearly show the great influence of the water content on the stiffness and on the permanent strain during the first 80 cycles of loading.

E_1 ranges from 500 to 1200 MPa. It can be modeled by (7).

$$E_1 = 452 - 174W + 97W^2 + 100F + 80M . \qquad (7)$$

The drier the material, the greater the stiffness. The stiffness also increases with the fillers content. Regarding material A, the stiffness is lower than for the material B. The aggregates of material B are more rigid and more angular than the material A, which prevents some of them from moving and cracking. The cumulated axial permanent strain can be modeled by (8).

$$\epsilon_1^{p^*} = 0,0049 + 0,0029W - 0,0019M - 0,0011W\text{-}M . \qquad (8)$$

$\epsilon_1^{p^*}$ is the axial permanent strain between cycle 2 and cycle 80 (9) to reduce the large scattering due to the first cycle.

$$\epsilon_1^{p^*}(N) = \epsilon_1^p(N) - \epsilon_1^p(N = 1) . \qquad (9)$$

The axial permanent strain increases with the water content. The strain level is greater for the material A than for the material B (for the same reasons as for E_1). The interaction between W and M shows that the influence of the water content is not the same for the two materials (the material A is more water-sensitive than the material B). Figures 6 and 7 represent the influence of the water content on the stiffness and the axial permanent strain for the two materials. If the water content increases by 1,5 % the stiffness of the specimens is divided by two and the axial permanent strain is multiplied by four. This will justify a further analysis on the influence of W on the strains under high number of cycles.

4 Conclusion

This first series of tests enables us to clearly show how the parameters of the initial state change when the nature of the material and the conditions of fabric vary. The dry density and the compaction strongly depend on the nature of the aggregates (fillers content, density and angularity) and on the intensity of compaction. The stiffness and the axial permanent strains during the first cycles of loading depend on the water content and the nature of the aggregates. The control of the initial state of the material on site (in particular the water content) is essential for the future behavior of the pavement. Actually a second series of tests is prepared with the purpose to relate the initial state (fabric and mechanical properties) with the long-term behavior of these materials.

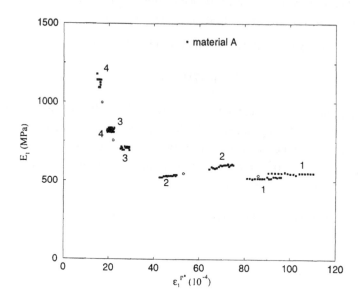

Figure 6: Influence of the water content on E_1 and on $\epsilon_1^{p^*}$ for the material A

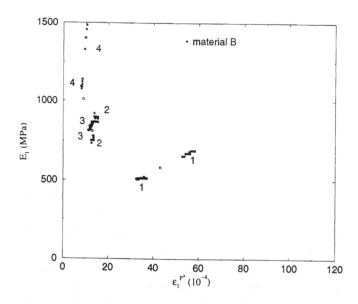

Figure 7: Influence of the water content on E_1 and on $\epsilon_1^{p^*}$ for the material B

5 Acknowledgements

The work presented has been carried out as a part of a research program supported by the "Laboratoire Central des Ponts et Chaussées". This support is gratefully acknowledged.

References

[1] SETRA-LCPC (1994): *Conception et dimensionnement des structures de chaussée.* Guide technique.

[2] PAUTE, J.-L., MARIGNIER, J., VIDAL, B. (1994): *Le triaxial chargements répétés LPC pour l'étude des graves non traitées.* Bulletin de Liaison des Laboratoires des Ponts et Chaussées, n°190, pp 19-26, Mars-Avril 1994.

[3] GIDEL, G., HORNYCH, P., CHAUVIN, J.-J., BREYSSE, D., DENIS, A. (1999-2000): *Nouvelle approche pour l'étude des déformations permanentes des graves non traitées lappareil triaxial chargements répétés.* Bulletin de Liaison des Laboratoires des Ponts et Chaussées, in press.

[4] LEKARP, F., RICHARDSON, I.-R. DAWSON, A. (199): *Influences on Permanent Deformation Behavior of Unbounded Granular Materials.* Transportation Research Record, n°1547, pp 68-75.

[5] PILLET, M. (1992): *Introduction aux plans d'expériences par la méthode de TAGUCHI.* Les éditions d'organisation, 224 p.

Soil compaction by kneading: comparison of existing methods and proposal of a new laboratory test

Hugues Girard[1], Denys Breysse[2], Richard Fabre[2], Paulin Kouassi[3], Daniel Poulain[1]

[1] Cemagref, 50 Avenue de Verdun, 33612 CESTAS cedex – France
[2] CDGA, Université de Bordeaux I, 33405 TALENCE cedex – France
[3] BNETD, Abidjan, Côte d'Ivoire

Abstract: In the majority of cases, fine soils used for the construction of earthworks are compacted by tamping rollers. It has been evidenced that the two methods usually employed in France for laboratory compaction, the normal Proctor and the static compaction test, do not always make it possible to obtain sufficiently representative specimens of this type of compaction *in situ*. Following an examination of the existing methods of kneading compaction, a new test is proposed. The physical and mechanical characteristics of specimens obtained in this test, called *"3-feet kneading"*, are compared to those of specimens taken from the site after compaction with a tamping roller. These comparisons evidence that the specimens obtained with the method as proposed are more representative of the soils compacted on site, even if this test can still be further improved.

1 Introduction

Fine soils used as backfill are generally compacted with tamping rollers. This form of compaction results in deformations in the soil in several directions as a result of the successive passages of the roller. This kneading action is different from other types of compaction, and results in a progressive change of the deviator ($\sigma_1 - \sigma_3$) at a given point in the soil. To foresee the behaviour of backfill in fine soils compacted in this way with sufficient accuracy at the study stage, better knowledge of the materials making up the soils is essential. Such an objective requires obtaining in the laboratory samples that are representative of the compaction on site, with densification methods as close as possible to the compacting physics of each type of machine. Such is not always the case with the usual laboratory compaction methods (normal Proctor compaction and static compaction). This finding is based on the experience of the Bordeaux regional centre of the Cemagref, which over the last 30 years has studied numerous dams in fine soils in South West France. Within the framework of these studies, using on site compaction tests areas, it has been recorded that the physical and mechanical characteristics of specimens of soils compacted in the laboratory with the usual methods differed significantly from those measured on specimens of soil compacted on site with tamping rollers (see figure 1). These observations are confirmed by the results presented in this paper (see chapters 4 and 5).

Figure 1: Tamping roller on site

The Cemagref has designed a new compaction method, called the "3-feet kneading compaction test" to obtain specimens that are more representative of the compaction of fine soils on site. The purpose of this new method is to reproduce, in the laboratory, the physical action of kneading implemented on site by tamping compactors. Compaction is performed in a CBR mould with a three-feet tool installed on a classic laboratory press. The design of this test is based, on the one hand, on the analysis of the different existing laboratory methods, and, on the other hand, on the examination of on site compaction conditions. The results obtained are examined by comparing different laboratory and on site compaction methods, also addressing the issue of the influence of the various compaction parameters.

2 Main existing kneading compaction methods

2.1 Background

The first record of work we were able to source on the theme of laboratory kneading compaction is very old, and indicates the concern of researchers with simulating the action of sheepsfoot compactors. As early as 1934, PHILIPPE (Zanesville District Office of the Engineers Corps) used a statically loaded, slow displacement punch. WILSON [12] tried to obtain laboratory curves ($\rho_d = f(w)$) close to site curves using a miniature HARVARD compaction test, where compaction was achieved using a hand hammer fitted with a spring. MCRAE and RUTLEDGE [8] described a static kneading compactor built at Northwestern University, with which it was possible to vary the load and contact time of the tamping tool with the material. By varying these two factors, the line of optima could be displaced, thus coming close to site conditions. The better known miniature HARVARD test and the tests developed more recently are described below. Table 1 shows the main characteristics of these trials.

Table 1: Summary of the main characteristics of compaction tests performed by kneading

Compaction	Laboratory				On site
	Harvard	Seed	Stojadinovic	Daoud	Tamping roller
Mould	$\varnothing$ = 3.3 cm h = 7.16 cm	$\varnothing$ = 10.16 cm h = 12.14 cm	H = 70 cm	Proctor	Layer of approx. 40 cm
Piston	$\varnothing$ = 1.27 cm	20.25 cm² Area of segment of the circle: ¼ of mould area	Tamper H = 18 cm pyramid frustum square b=5cm	$\varnothing$ = 2 cm	Tamping foot
Pressure (MPa)	0.70 to 2.46	2.07	13.5	1.6	Variable
No. of layers	5	5	2	5	-
Number of applications of load per layer	10	25	8	2 x 25	Generally between 6 and 10
Application time	15 seconds for 10 applications	30 applications a minute	-	30 applicat. a minute	< 1 second
Criterion for stopping an application	Pressure	Pressure	Sinking	Pressure	Pressure and sinking

2.2 HARVARD miniature compaction

This method was developed at HARVARD University [12]. Compaction is performed manually in 5 layers in a mould 7.16 cm high and 3.30 cm in diameter. The piston has a diameter of 1.27 cm. Compaction pressure is controlled by a spring set at 1.405 MPa. A total of 10 compressions are made per layer, applied in 15 seconds.

The small size of the mould used limits the application of this test to soils with a very small particle size. No large size specimens can be obtained.

2.3 Work of RASDSCHELDERS

This work, dating from 1953, is quoted by STOJADINOVIC [11]. The large-size device aims at reproducing on site compaction conditions using sheepsfoot rollers. The soil to be compacted is placed in a metal tray up to 70 cm high, installed on a mobile trolley enabling the tamper to be applied to any point on the surface of the soil. The tamper is a pyramid frustum, 18 cm high, with a 5 cm square base. The load on the foot is transmitted by means of a lever. The driving force of the foot is adjusted by means of a winch system.

2.4 Triaxial Institute Kneading Compactor

This device, proposed by SEED and MONISMITH [10], was automated so as to control the pressure exerted by the foot and the application time of the pressure. The specimens are 15.24 cm in diameter and 30.48 cm high. The foot is in the shape of a disk segment with a radius equal to that of the mould and a surface area equal to one quarter of the cross section of the mould. SEED [9] studied in particular the influence of the depth of penetration of the foot on soil resistance.

2.5 Kneading test with the DAOUD cylinder [1996]

In parallel to the trials performed by the Cemagref to study the influence of kneading compaction on soil mechanical behaviour, similar work was carried out at the ENSG in Nancy to examine the consequences of this type of compaction on soil permeability.

DAOUD [3] developed a kneading method drawing direct inspiration from the HARVARD test, but using the normal Proctor mould. Compaction was performed using an automated device, in 5 layers, with 2 x 25 applications by layer of a piston 2 cm in diameter at a rate of 30 applications a minute. Compaction pressure was 1.6 MPa, the objective being to obtain the same densities as with the HARVARD test.

2.6 Summary of the laboratory kneading trials and evolution of on site compaction machines

The methods presented correspond to sheepsfoot compactors. These machines have small feet so as to compact by the application of a high normal stress.

These machines have now been replaced in the majority of cases by tamping rollers, which are more efficient for the compaction of fine soils with a water content close to the normal Proctor optimal water content. In comparison to sheep feet, the tamping feet have a larger contact surface area and are parallelepiped in shape, offering less resistance to forward movement and enabling the ramming action to be complemented by kneading.

This evolution in the design of compaction machines encouraged us to propose a new method, making it possible to take account, in particular, of pressure and driving as criteria for stopping the application of the load.

3 Test proposed: 3 feet kneading tool

3.1 Principles of design

The aim is to represent as closely as possible the action of a tamping roller on site. The machine taken as a reference is the CATERPILLAR 825 C.

To define the laboratory test, the following proportionality ratios were taken:
- the proportionality ratio between the total surface area of a wheel and the surface area actually compacted (surface area of the feet in contact with the soil): ratio of 3.7;
- the proportionality ratio between the thickness of the compacted layers and the height of the tamping feet : ratio of 1.5.

The tool had to enable stopping of the driving of the feet either on reaching a maximal constraint under the feet, or a global force corresponding to total penetration of the feet. The aim is to simulate the action of the roller which, in the initial passes or in the case of wet soil, sinks down to the generatrix of the wheels, and which in subsequent passes rests on the feet.

The maximal stresses applied under the roller feet on site are not precisely known. The values given in the literature vary considerably, from 1 to 5 MPa. We attempted to measure these stresses on site using sensors, but measurements proved to be extremely delicate. In particular, the positioning of the sensors in relation to the feet was very difficult. Nevertheless, values ranging from 3 to 5 MPa were measured for an all but static application of the constraint (machine moving extremely slowly). The test must simulate a constraint value and application conditions (speed and duration) that are properly representative.

Furthermore, it had to be possible to perform the test with the maximum number of the types of equipment currently available, in a classic soil mechanics laboratory. The size of the specimens obtained had to be sufficient.

3.2 Equipment adopted

The equipment adopted, in addition to the three-feet tool (see figure 2), consists of

classic equipment for geomaterial laboratories :
- a CBR mould,
- a press for triaxial compression tests,
- the three-feet compaction tool.

The three-feet tool proposed comes in the form of a 15 cm diameter steel disk, under which three steel cylinders, 5 cm in diameter and 2.5 cm high, are fitted. The CBR mould has an inside diameter of 15.2 cm and a height of 15.2 cm.

Figure 2: Presentation of the three-feet kneading tool

Use of the CBR standardised mould is explained by the desire to obtain samples of representative size. Several specimens can be taken from the same mould for different tests, for a possible study of permeability, for the study of the soil shear resistance, etc. Such an arrangement offers the advantage of ensuring the same initial physical conditions for the various specimens, thus making the conclusions more reliable. Edge effects are usually attenuated in comparison with smaller moulds.

3.3 Main points of the test procedure

Samples are prepared in accordance with the NF P94-093 standard for the normal Proctor test, to make the results comparable, especially as far as the physical properties of the compacted soil are concerned.

Several compaction pressures have been tested using this process [5]. The first tests were performed with compaction pressures of 1 MPa, 1.25 MPa and 1.5 MPa respectively in 3 layers and in 5 layers. On the materials studied, the dry weights closest to the results obtained on site are those from 5 layer compaction with a pressure of 1.25 MPa. On examining the results more closely (see Chapter 5), it can be seen that this density criterion is insufficient and that at equal density an increase in the stress makes it possible to obtain mechanical characteristics which are more representative of on site compaction.

Force is applied by a metal rod, or piston, integral with the disk and centred in relation to the three feet, forming an equilateral triangle (see figure 2). The effort applied can thus be considered as being equally spread over each of the three feet. The rigidity of the rod-disk-feet system is such that the 3-feet tool cannot be tilted during compaction. The tool is displaced at a constant speed.

The arrangement of the three feet is such that they have to be applied eight times to compact each of the 5 layers in a uniform manner. That corresponds to 8 passes, an order of magnitude generally practised on site.

The first drive of the feet, or the first pass, produces compaction of approximately one third of the surface of the mould. The depth to which the feet are driven is of no concern to the operator as it is the compaction effort, representing the weight of the machine, that is used as a criterion to stop driving. Depending on the nature of the material and on its water content, therefore, stoppage occurs either before the plate comes into contact with the soil (the force applied being supported solely by the feet) or when the feet have been driven in completely, the force applied then being spread over the entire section of the mould through the plate. This situation is identical to the behaviour of the tamping feet on site, as described in chapter 3.1.

When the targeted force is reached, the feet are withdrawn to be applied once again after a rotation of the mould through $1/8^{th}$ of its circumference.

3.4 Homogeneity of the specimens compacted with the 3-feet tool

This criterion is, of course, the first to be verified for validation of the method proposed. The homogeneity of the specimens obtained was examined on a silty clay. Three water contents were chosen for this study ($W_{OPN}-2$, W_{OPT} and $W_{OPN}+3$). The compacted samples ($\varnothing$ 15.2 cm, h = 15.2 cm) were cut after removal from the mould into four horizontal slices to obtain measurements of the dry density of the soil including the top and bottom parts of each layer together with the interface of two successive layers.

Figure 3 presents the profiles of the densities obtained in the case of compaction with the normal optimal Proctor water content (W_{OPT}). These profiles evidence the good homogeneity of the densities measured, horizontally and vertically. For the three water contents tested, the variation coefficient does not exceed 1.4% in the worst case (W_{OPN}-2).

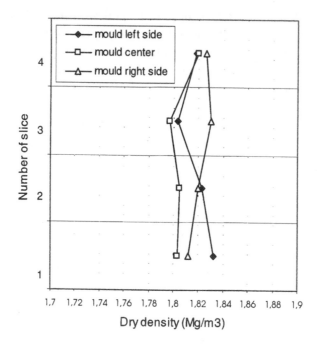

Figure 3: Example of homogeneity study result
(compaction pressure: 1.25 Mpa; W_{OPT})

4 Results obtained

4.1 Comparison on the laboratory compaction tests

A comparative study of kneading compaction methods was performed on the Xeuilley silt used by DAOUD [3] in his work at the ENSG in Nancy. Two samples, Xeuilley_1 and Xeuilley_2, were used respectively at the ENSG by DAOUD [3] and at the Cemagref by KOUASSI [6].

The results shown in figure 4 enable comparison between kneading compaction and the normal Proctor compaction. The same and equal compaction pressure of 1.6 MPa was used for the 3 kneading methods. It can be seen that these 3 tests resulted in optimum dry soil density higher than those obtained with the normal

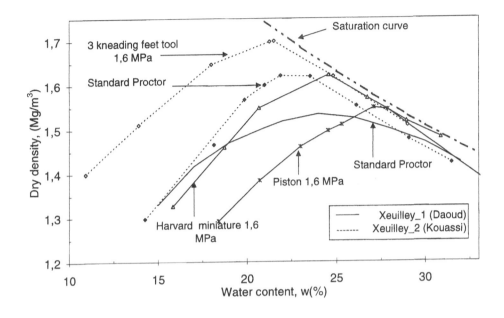

Figure 4: Compaction curves of Xeuilley silt

Proctor test. The optimal water content is slightly lower than or equal to W_{OPT} for the tests with the 3-feet tool and the miniature Harvard, but considerably higher for the cylinder kneading developed by DAOUD. It can also be seen that the wet branch of the 2 curves for the miniature HARVARD compaction and the 3-feet tool is much closer to the saturation curve than the Proctor curves.

4.2 Comparison with on site compaction using tamping roller

Figure 5 shows the results of the density control performed on a layer of material during the construction of the Fargues dam and the compaction tests performed in the laboratory on the same soil. It can be seen that for this silty clay compaction with the 3-feet tool is much more representative of the dry density obtained with the tamping roller. In particular, on the wet side of the optimum, the degrees of saturation obtained by normal Proctor compaction, hitherto the test of reference for the control of compaction of fine soils, are lower than those measured on site.

5 Influence of the mode of compaction on the mechanical characteristics of fine soils

It has been seen above that the laboratory compaction method using the three-feet

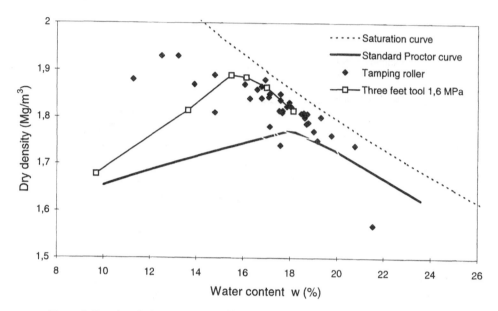

Figure 5: Density of Fargues silty clay obtained with different compaction methods

tool made it possible to obtain homogenous samples of sufficient size to perform mechanical tests and dry densities close to those obtained on site with tamping roller. The following part will look at the mechanical properties of the laboratory compacted soils and the representativeness of those properties in relation to those of fine soils compacted on site.

5.1 Behaviour under slight deformation

Triaxial unconsolidated and undrained tests (UU) were performed on samples from one and the same soil from the Fargues dam, compacted according to the following methods:

- on site with a tamping roller; intact 20 cm cubes were taken from the dam and the tri-axial specimens made from the cubes;
- by static compaction in 3 layers in a mould the size of the specimen (35 mm in diameter), the method generally used in France for the study of fine soils;
- by kneading with the 3-feet tool in a CBR mould; two tests with pressures of 1.25 and of 3 MPA respectively were performed.

The results of the triaxial tests performed on the specimens compacted as above are shown in figures 6 and 7, corresponding respectively to confinement pressures, σ_3, of 100 kPa and 400 kPa, which are the values generally taken for the study of medium height backfills (from 10 to 20 m high). These figures evidence very

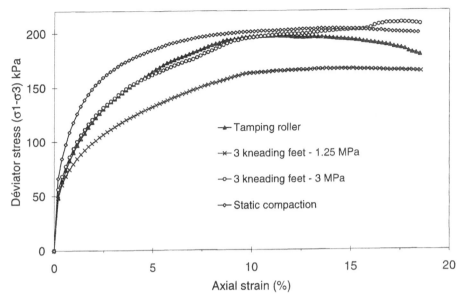

Figure 6: Stress-strain curves (UU test for $\sigma_3 = 100$ kPa)

different types of behaviour with slight deformation according to the compaction methods used.

At the low confinement pressure, σ_3 equal to 100 kPa, it can be clearly seen that compaction with the three-feet tool at a pressure of 3 MPa results in a mechanical behaviour very close to that of the soil compacted on site, the stress-strain curves being practically identical. The resistance of the soil compacted statically presents, with low strain, a resistance greater than that obtained with the two other modes of compaction. In particular, its initial tangential modulus is much greater.

With a higher confinement pressure ($\sigma_3 = 400$ kPa), the influence of the compaction pressure with the 3-feet tool is less marked. Irrespective of the compaction pressure (1.25 MPa or 3 MPa), the effort-deformation curves are close to the curve obtained with compaction on site. As in the case where the constraint, σ_3, is equal to 100 kPA, it can be seen that the statically compacted specimen has a stiffness greater than that obtained with the other compaction methods.

Other tests of the same type performed on other fine soils from engineering structures in South West France confirm that static compaction results in an overestimation of the initial tangential modulus of the soils compacted. On the other hand, the laboratory kneading method proposed makes it possible to reproduce specimens which are representative of the low deformation behaviour of fine soils compacted with tamping roller.

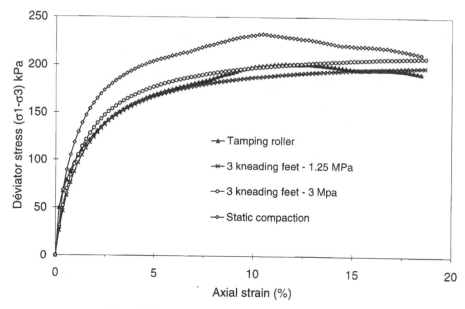

Figure 7: Stress-strain curves (UU tests for $\sigma_3 = 400$ kPa)

5.2 Limit state parameters

Contrary to cases of low deformation, it can be seen from Figure 7 that at the point of breakage the stress-strain curves are very close for all methods of compaction. Figure 8 shows the influence of the compaction methods on the undrained cohesion *Cuu* of the silty clay of Fargues. Each test consists of the triaxial compression of 4 specimens ($\sigma_3 = 50, 100, 200$ and 400 kPa) and the scattering of the results is represented by the segments in the graph.

The levels of cohesion obtained by on site compaction, by laboratory kneading (3 MPa, 8 mm/min) and by static compaction (static 2) are comparable. It should be pointed out that the uncertainties as to the value of the cohesion of the specimens compacted on site can be explained by the variability of the soil, which is not homogenised on site as it is in the laboratory. The scatter observed with the 3-feet tool (3 MPa) is related to the buckling of certain specimens in the course of the triaxial test.

For the statically compacted specimens, cohesion is closely related to the dry density of the specimens. With 3-feet tool kneading, the three modes of operation resulted in the same dry density of the soil, but cohesion increases with the pressure applied and when the drive speed decreases.

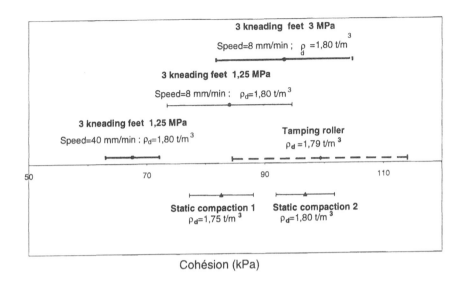

Figure 8: Undrained cohesion of Fargues silty clay (UU tests)

Under such conditions, it would appear that for the silty clay of Fargues kneading compaction using the 3-feet tool with a pressure of 3 MPa and a speed of 8 mm/min makes it possible to reconstitute, in the laboratory, specimens which are representative of the mechanical behaviour of the backfill on site, both for slight deformations and for limit state parameters.

6 Conclusion

This work made it possible:
- to develop and exploit a new method, more representative of on site compaction conditions using tamping roller (the feet sink less into the soil and binding of the soil is improved between the three feet) ;
- to evidence a relationship between the soil deformation modulus and the mode of compaction.

Examination of the results obtained with this new method has evidenced that :
- the dry density ρ_d from kneading is more representative of the ρ_d on site than the normal Proctor ;
- kneading compaction results in :
  - a dry density ρ_d greater than that obtained with the normal Proctor test, a difference that is considerable on the dry side of the OPN,
  - a high degree of saturation on the wet side of the compaction curve.

- static compaction results in an overestimation of the real stiffness of the soil ;
- kneading compaction makes it possible to represent better the mechanical behaviour of the soil, both in low deformation (initial tangential modulus) and at limit state (Mohr-Coulomb criterion) ;
- the joint study of the physical and mechanical properties of soils compacted by kneading evidences the limits of a compaction control based solely on the ρ_d when the kneading effect is predominant.

Our study into the kneading compaction of fine soils will be pursued to clarify the following points :

- measurement of compaction pressures on site, under the feet of the tamping rollers, deserves to be pursued, taking account of the lessons drawn from the measurements recorded ;
- the influence of the speed and pressure of laboratory kneading compaction was evidenced within the ranges of the variations limited by the use of classic presses. Analysis of this influence must be extended to higher ranges of speeds and pressures ;
- the kneading compaction method that we have developed needs to be validated on a greater number of soils, in particular taking into account the role of mineralogy.

The authors acknowledge ENSG Nancy and in particular F. MASROURI who permits the comparison of compaction methods on Xeuilley silt.

References

[1] AFNOR NFP-94093. (1995): *Géotechnique, vol. 1: essais de reconnaissance des sols (frenchnorm: field tests for soils)*.

[2] CAMAPUM DE CARVALHO. J. (1985): *Problèmes de reconstitution des éprouvettes de sol en laboratoire; théorie et pratique opératoire (methodology proposal for rebuilding soil specimens in laboratory)*. LCPC (Central Highways Laboratory) research report. 145. 54 p.

[3] DAOUD, F. (1996): *La perméabilité des sols fins compactés (permeability of compacted clays)*. PhD Thesis I.N.P.L. Ecole Nationale Supérieure de Géologie de Nancy, 194 p.

[4] DUNCAN. J.M., PETER, B., et al. (1980): *Strength, stress-strain and bulk modulus parameters for finite element analyses of stresses and movements in soil masses*. College of Engineering, Office of Research Services, University of California, Berkeley. 70 p+ appendix.

[5] GRAZIANA, J.M. (1994): *Compactage par pétrissage de sols fins en laboratoire (kneading compaction of clays in the laboratory)*. Eng. report, CUST de Clermont-Ferrand, 123 p.

[6] KOUASSI P. (1998): *Comportement des sols fins compactés: application aux remblais et aux ouvrages en terre. (behavior of compacted fine soils: application on embankments)*. PhD Thesis. Université de Bordeaux I. 174 p.+ appendix.

[7] LEONARDS, G.A. (1968): *Les fondations*. Dunod. 1106 p.

[8] MCRAE, J.L., RUTLEDGE, P.C. (1952): *Laboratory kneading of soil to simulate field compaction.* Proceedings Highway Research Board, Vol. 31, pp. 592-600.

[9] SEED, H.B., CHAN, C.K. (1959): *Structure and strength characteristics of compacted clays.* Highway Research Board Bull, P 87-128.

[10] SEED, H.B., MONISMITH, C.L. (1954): *Relationship between density and stability of subgrade soils.* Highway Research Board Bull, pp. 16-32.

[11] STOJADINOVIC, M.R. (1964): *Etude des conditions de compactage des sols avec les rouleaux à pieds de mouton et les rouleaux à pneus (study of compacting conditions with sheepsfoot compactors and tire compactors).* Annales. I.T.B.T.P., 195-196. Mars-Avril 1964. pp.301-334.

[12] WILSON, S.D. (1950): *Small soil compaction apparatus duplicates field results closely.* Engineering News-Record, Vol. 145, 18, pp. 34-36.

[13] WILSON, S.D. (1970): *Suggested method of test for moisture-density relations of using Harvard compaction apparatus.* Special procedures for testing soil and rock for engineering purpose. ASTM STP 479.

Soil Compaction by Cyclic Loading: Ins and Outs

Alain Vuez, Christophe Lanos, Abdou Rahal

Groupe de Recherche Génie Civil Rennes, Laboratoire Matériaux, INSA de Rennes, France

Abstract: Clay soils are well known in Civil Engineering for their capability to generate excess pore pressure when loaded. As shear strength and settlements are widely influenced by the range of the excess pore pressure and its dissipation, its prediction reveals itself very important.
Cyclic loading is a special case encountered in different and various situations as floods, tide or filling and emptiing of a silo. All these loadings are function of time, have low frequency and often conduct to a stationary state.
This paper deals with the excess pore pressure: its amplitude, its out of phase, its relation with soil parameters, geometry of the problem and loading conditions. An analytical solution is proposed, validated by experiments.
Principal results are to explain the conditions for which amplitude of excess pore pressure is upper than amplitude of loading, and to discuss the importance of compressibility of pore fluid by comparison with compressibility of soil.

Introduction

A loading at the boundary of a clay layer creates an excess pore pressure within the layer, that is at the origin of settlements and lack of stability. So firstly to quantify the level of excess pore pressure and secondly its action are not so easy and new ways are still opened to describe and analyse the phenomenon.

Settlements are explained by the consolidation process with dissipation of pore pressure when occurs volume variation of the soil skeleton. Process is a transport problem coupled with conservative equation. Fluid flow is generaly governed by Darcy's law.

Usually, problem is solved with two simplifying hypotheses:
i   initial condition such that excess pore presure is equal to the load increase,
ii   drainage boundaries conditions are zero on the both sides of the soil layer, or of the soil sample in an oedometer test (usualy considered for this problem).

The Terzaghi's solution is the most usefull one in practical engineering. But comparison with the reality shows immediatly that loading is not instantaneous, and there is no discontinuity of the pore pressure at the boaders of the clay layer from the beginning to the end of the consolidation process. In the following is developped a solution that takes into account these remarks. Initial condition is zero as drainage condition. But to do not have a trivial solution, it is necessary to introduce loading function of time.

Construction of buildings, embankments or dams are described by a constant rate of loading « CRL » already treated by [1]. Cyclic loading is an another kind of loadings function of time, often encountered with variations of water levels, floods, waves, or tide. A special case of loading at the surface of a clay layer is the filling and emptying of a silo.

Caracteristic of these cyclic loadings is their large period. That permits to study them as a consolidation process and not as a propagation process. By that way the consolidation equation is set using a sinusoidal loading, and compressibility of pore fluid is considered as well as compressibility of soil skeleton. A solution is proposed, validated by experiments with the oedometer test on clay.

The expected conclusions are to investigate about the ratio of the excess pore pressure on the increase of loading, inside a layer, and the role of the compressibility of pore fluid.

1   Equation of consolidation process

Consolidation process is discribed as a diffusion phenomenon where continuity equation of fluid is written with the use of Darcy law. In monodimensional condition, one obtains [2] and [5] the equation:

$$\frac{\partial \varepsilon_v}{\partial t} + \frac{n}{\rho_w} \frac{\partial \rho_w}{\partial t} = \frac{1}{\rho_w g} \frac{\partial}{\partial z}(k \frac{\partial u}{\partial z}) \ . \tag{1}$$

where ε_v represents volumetric strain of soil matrix,

n porosity,

ρ_w unit weight of water,

γ gravity,

k permeability,

u pore water pressure.

The consolidation equation is solved with pore pressure u as main variable. Three more equations are necessary : one state equation for soil skeleton, one for pore fluid and one equation binding effective stress σ' and applied stress σ:

$\varepsilon_v = m_v \Delta \sigma'$ with m_v compressibility coefficient of skeleton,

$d\rho_w = - \rho_w \beta du$ with β compressibility coefficient of pore fluid,

$\sigma' = \sigma - u$ Terzaghi's postulate

Equation (1) is written as, [4]:

$$\frac{1}{c_v}\frac{\partial u}{\partial t} = \frac{\partial^2 u}{\partial z^2} + \frac{1}{c_{vg}}\frac{\partial \sigma}{\partial t} \qquad (2)$$

where consolidation coefficients c_v and c_{vg} are given by:

$$c_v = \frac{k}{\rho_\omega \gamma(n\beta + m_v)} \qquad \text{and} \qquad c_{vg} = \frac{k}{\rho_\omega \gamma m_v} \qquad (3)$$

The proposed solution of (2) corresponds to the real conditions of a layer of height H, drained on one side. Such a condition is encountered in the oedometer test with pore pressure measurement. One have:

        initial condition:

    pore pressure u constant with respect to depth z $u(z,0) = u_0$.

        boundaries conditions:

    drainage condition with controled pressure at the top ($z = H$): $u(H,t) = f(t)$.

    and impervious boundary at the bottom ($z = 0$): $\dfrac{\partial u}{\partial z}(0,t) = 0$

Function $\dfrac{\partial \sigma}{\partial t}$ is called $g(z,t)$. Then one obtains the solution:

$$u(z,t) = \frac{2}{H}\sum_{m=1}^{\infty}\cos\frac{Mz}{H}\, e^{-M^2 T_v}\int_0^H \cos\frac{Mz}{H}\, u_0\, dz$$

$$+ \frac{c_v}{c_{vg}}\frac{2}{H}\sum_{m=1}^{\infty}\cos\frac{Mz}{H}\, e^{-M^2 T_v}\int_0^t e^{-M^2 T'v}\int_0^H \cos\frac{Mz}{H}\, g(z,t')\, dz\, dt'$$

$$+ c_v\frac{2}{H^2}\sum_{m=1}^{\infty}\cos\frac{Mz}{H}\, e^{-M^2 T_v}\, M \sin M \int_0^t e^{-M^2 T'v} f(t')\, dt' . \qquad (4)$$

with $M = (2m-1)\dfrac{\pi}{2}$.   and $T_v = \dfrac{c_v t}{H^2}$.   (5)

1.1   Sinusoidal loading

Initial and boundaries conditions for this type of loading are written as :

$$\sigma = Ap \sin \omega t + \sigma_0 . \qquad \text{and} \qquad g(z,t) = \frac{\partial \sigma}{\partial t} = Ap\, \omega \cos \omega t .$$

$$u_0(z,0) = u_0 \qquad ; \qquad u(H,t) = f(t) = u_0 .$$

The solution of (4) is [5]:

$$u(z,t) = u_0 +$$

$$2Ap \, (n_{c_f} - 1) \sum_{m=1}^{m=\infty} \frac{(-1)^m}{M + \theta_v^2 M^5} \cos \frac{Mz}{H} (\sin\omega t + M^2 \theta_v \cos\omega t - M^2 \theta_v \, e^{-M^2 T_v}) \qquad (6)$$

with $\qquad \theta_v = \dfrac{c_v}{\omega H^2} \qquad$ and $\qquad n_{c_f} = \dfrac{n\beta}{m_v + n\beta} = 1 - \dfrac{c_v}{c_{vg}}$

θ_v coefficient is similar to time factor T_v. It depends of the quality of soil and of pore fluid, the thickness of soil layer and the angular frequency of loading.

When a stationary state is reached, the equation of pore pressure u_b at $z = 0$ (impervious end of the layer or sample) is:

$$u_b = u_0 + Ap_{ub} \sin (\omega t + \phi_{ub}) . \qquad (7)$$

The amplitude Ap_{ub} of pore pressure u_b is died down with respect to loading Ap. The attenuation coefficient is defined by:

$$Att_{ub} = \frac{Ap_{ub}}{Ap} .$$

Phase lag of pore pressure u_b with respect to loading is expressed by $Deph_{ub}$ measured in time unit or as well by a non-dimensional term, if T represents the period of loading:

$$DéphT_{ub} = \frac{Déph_{ub}}{T} \quad \text{where} \quad \phi_{ub} = 2\pi \, DéphT_{ub} \qquad (\phi_{ub} \text{ in radian unit})$$

The drawing of curves showing attenuation coefficient with respect to θ_v within different rates of compressibility coefficient n_{cf} is of interest. Figure 1 gives the curve for $n_{cf} = 0$ (incompressible fluid). Variations for others values of n_{cf} may be obtained by an affinity of ratio $(1 - n_{cf})$. Zero value of n_{cf} represents the special case of an uncompressible fluid. This kind of hypothesis is often used in soil mechanics design.

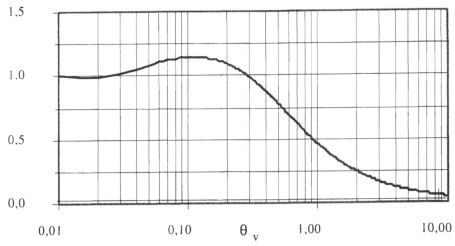

Figure 1: Attenuation versus θ_v - Sinusoidal loading

Two comments may be done :

1. amplitude of pore pressure may overflow amplitude of loading for values of θ_v in the range 0.03 to 0.3,

2. pore pressure u_b admits an out of phase forward with respect to the loading from a value of θ_v equal to 0.05.

Figure 2 represents out of phase variation with respect to θ_v. It does not depend of n_{cf}

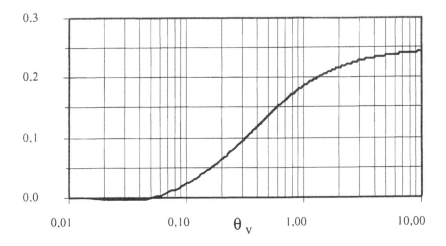

Figure 2: Out of phase of pore pressure u_b

The first observation is analysed in the following by a sketch of isochrones lines of pore pressure. Experimental data obtained by oedometric tests confirm such variations.

2   Experiments

Experimental study has been worked out by means of an apparatus described below.

2.1   Experimental apparatus

A sample is enclosed inside a metal ring to ensure no lateral stains. Measurement of pore pressure is performed on one side of the sample (at the bottom of the sample). On the other side (at the top), a drainage condition is imposed with controled pressure by mean of a porous plate. Pressure at the top of the sample acts as back pressure before loading and during the equilibrium state for the initial condition. A controled pressure with respect to time may be performed during a test when drainage condition is related to time, and represented by the f(t) function in the solution (4).

A perfect waterproof between sample and metal ring must be achieved. Vertical load is applied by a jack and vertical displacement is measured following the head jack displacement.

Data acquisition and control system (« Keithley 570 ») are joined to a personal computer. Some calibrations and experimental errors had have to be first quantified such as :
- stains of the cell itself
- delay of pressure measurement because of load effect upon the sensor : needed variation of volume to ensure deformation of sensor membrane.

2.2   Soil caracteristics of samples

The tested soil has been extracted from a place called « Claye Souilly » and is a silt traited with bentonite. Two samples have been cut and their caracteristics are set on the table I.

Sinusoidal loadings have been carried out with different angular frequencies within the range of 0.105 to 8.7 10-5 Hz. (for a period from 1mn to 20 hrs). The amplitude has been chosen 25 kPa : a level considered optimum as regards measurement of attenuation and less influence of the overconsolidation during process.

Table 1: Caracteristics of samples

	Unit	Sample 3	Sample 5
Unit weight	kN/m^3	20.85	21.59
Solid unit weight	kN/m^3	26.4	26.4
Initial water content	%	11.2	15.0
Initial void index		0.410	0.408
Blue test value		2.4	-

2.3   Results

Table 2 resumes the different loadings and the obtained values of the coefficient θ_v, the consolidation coefficient c_v and n_{ct}

For each stage, the consolidation coefficient c_v is calculated by means of the out of phase of pore pessure. It is not linked with the coefficient n_{ct}. Knowing c_v and measuring attenuation, the coefficient n_{ct} is calculated.

Table 2: Loadings and coefficiens

Sample	Stage	Hight (m)	Void ratio	Period (s)	*Attu*	*Deph*	θ_v	c_v (m^2/s)	n_{ct}
Claye 3	310/4	0.0112	0.340	60	1.126	0.03	0.115	1.5e-06	0.018
Claye 5	501/5	0.0103	0.319	60	0.941	-0.075	0.050	5.6e-07	0.227
Claye 5	501/21	0.0100	0.282	300	1.042	0.0	0.050	1.0e-07	0.039
Claye 5	501/23	0.0100	0.283	900	1.052	0.0	0.050	3.5e-08	0.030
Claye 5	502/2	0.0101	0.290	1800	1.079	0.023	0.100	3.5e-08	0.059
Claye 5	502/3	0.0101	0.289	7200	0.874	0.077	0.238	2.1e-08	0.170
Claye 5	503/2	0.0101	0.291	1200	1.091	0.012	0.075	4.0e-08	0.036
Claye 5	504/2	0.0101	0.293	3600	1.003	0.079	0.215	3.8e-08	0.071
Claye 5	504/4	0.0101	0.293	5400	0.915	0.086	0.266	3.2e-08	0.106
Claye 5	505/2	0.0101	0.293	10800	0.764	0.105	0.339	2.0e-08	0.189
Claye 5	506/2	0.0101	0.294	9000	0.803	0.100	0.315	2.2e-08	0.172
Claye 5	506/3	0.0101	0.294	18000	0.573	0.129	0.450	1.6e-08	0.305
Claye 5	507/3	0.0101	0.296	7200	0.889	0.083	0.257	2.3e-08	0.141
Claye 5	508/2	0.0101	0.291	36000	0.385	0.157	0.635	1.1e-08	0.419
Claye 5	508/3	0.0101	0.293	72000	0.218	0.192	1.100	1.0e-08	0.489
Claye 5	509/2	0.0101	0.294	54000	0.284	0.172	0.785	0.9e-08	0.489

2.3.1   Attenuation higher than unity

The attenuation is higher than unity for six loadings, with an overtaking that can reach 13 %. That means an over pressure by comparison with total stress loading. This situation is provided by the theory (see figure 1), and appears with Clay sample 3 for a period of 60s (see figure 3) and Clay sample 5 for a period 1800s (see figure 4). Both periods are not equal for samples have different consolidation coefficient.

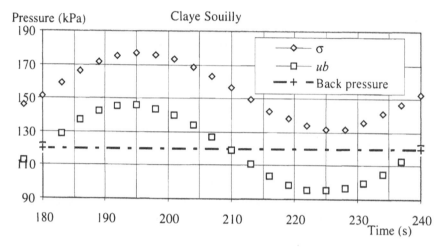

Figure 3: Loading and excess pore pressure - Claye Souilly sample 3 / stage 310/4

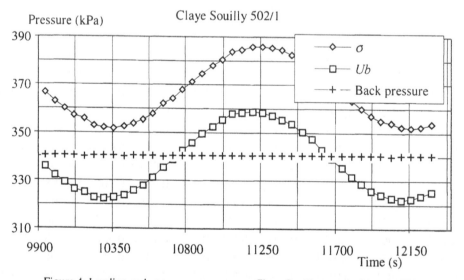

Figure 4: Loading and excess pore pressure - Claye Souilly sample 5 / stage 502/1

Experimental results of stage 310/4 have been compared with calculated results by (6) with coefficient θ_v, = 0.11 and n_{cf} = 0.018. Figure 5 shows they are very closed. The use of (6) permits the calculus of pore pressure at different depths and for different times

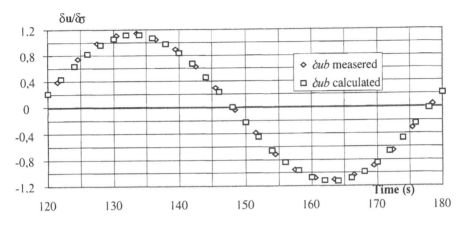

Figure. 5: Comparaison between measured and calculated excess pore pressure
Claye Souilly 3 / stage 310/4

One can see the drawing of isochrones of pore pressure in the figure 6. Inspection of the values of the ratio $\Delta u/\Delta\sigma$ is interesting and show that its optimum is always located at $z/H = 0$ but sometimes inside of the sample. For instance, the ratio is upper than unity at z/H equal to 0.7 and 0.6 for $t = 58s$ and 60s respectively.

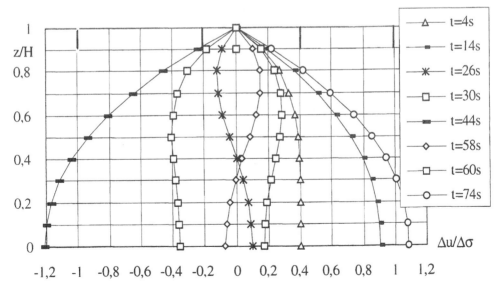

Figure 6: Isochrones of excess pore pressure within the sample for different time

2.3.2   Study of n_{cf} coefficient

When considering fluid as incompressible (hypothesis of Terzaghi's solution), coefficient n_{cf} is equal to zero. Experimental values of n_{cf} are quite different and increase as the period of loading increase. Using measurement of settlement during the test, determination of the skeleton compressibility m_v of soil and the compressibility of water β is possible [5]. Figure 7 shows the variation of m_v and β. Compressibility of soil is constant with respect to the period but β is increasing with period. This fact is amazing and not explain at this day.

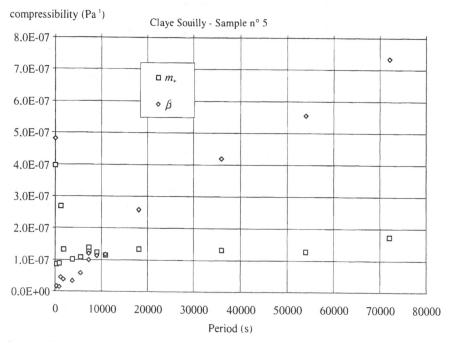

Figure 7: Variation of compressibility of skeleton m_v and of pore fluid β versus period - Claye Souilly sample n°5

Conclusion

Excess pore pressure is an important factor when studying the economical consequences on stability or settlement of structures or embankments. Duration of consolidation process is generally calculated using the simplest hypothesis where load is applied by a single step. In most cases, it induces no features on earth stability. However load step is unrealistic.

A loading with respect to time is more representative of real loading. There is no discontinuity between initial condition and boundaries conditions. Cyclic loading is of particular interest because it appears phenomenon of attenuation, amplification or filter. A consolidation process (diffusion process) is sufficient to explain when excess pore pressure can be larger than those predicted by step loadings.

A general solution is proposed in this paper, and the special case of sinusoidal loading is developped as example. Theoritical results are confirmed by experiments with an oedometer, using samples of mixed silt-bentonite soil. It has been encountered that compressibility of pore fluid reaches compressibility of skeleton, and that pore pressure exceeds loading more than 15%, in some cases when angular frequency, thickness of soil layer and consolidation coefficient are fited.

Cyclic loadings appear in different conditions as waves, tide, floods or filling of a silo [3] and [6]. Period is in the range af some hours to a year, but the thickness of the layer may also varied in a large scale. Within a stationary state, excess pore pressure does not vanish and can be harmful to stability of earth constructions or river banks.

References

|1| ABOSHI H., YOSHIKUMI H. et MURAYAMAS S. - *Constant loading rate consolidation test*. Soils and Fondation. Vol. 10. N°1 Pp. 43-56 (1970).

|2| BERRY P.L. and POSKJTT T.J. (1972): *The consolidation of peat*. Geotechnique. Vol. 22, 27-52

|3| FAVARETII M. and MAZZUCATO A. , (1994*) Settlements of a silo subjected to cyclic loading*. Proceedings of Settlement 94 , Pub. N°40 ASCE, Vol 1, 775-785 (16-18 Juin 1994).

|4| OZISIK N. (1980): Heat conduction, John Wiley.

|5| RAHAL A. (1993) *Etude de la consolidation unidimensionnelle d'un kaolin soumis à des chargements par paliers et sinusoïdaux*. Thèse I.N.S.A. de Rennes

|6| RAHAL A. and VUEZ A. (1998) *Analysis of settlement and pore pressure induced by cyclic loading of a silo*. Journal of Geotechnic Engineering ASCE, vol. 124, issue 12, 1208-1210.

|7| VUEZ A. and RAHAL A. (1994*) Cyclic loading for the measuring of soil consolidation parameters*. Proceedings of Settlement 94 , Pub. N°40 ASCE, Vol 1,. 760-774 (16-18 Juin 1994).

Interpretation method of compaction test results on powders

Christophe Lanos, Alain Vuez

Groupe de Recherche Génie Civil Rennes, Laboratoire Matériaux, INSA Rennes, France

Abstract : Exploitation of the analytical solution written in the case of simple compression test with perfect plastic fluid leads to an identification method of the yield criterium corresponding to pastes or pulverulent products. A such identification method has been assessed and validated by the use with ceramic paste, sand or dry clay powder. By considering the analogy between compression test and pressing test, characterization of powder behaviour is made possible thanks to compression tests with confinement. Pressing test data on cellulose powders show the benefits of the procedure.

1   Introduction

Industrial processes are the result of combining elementary actions that are often not always so easy to distinguish. The manufacturing process control is often realised during production, with the help of easy tests or records of some manufacturing data. Control is based on the analysis of the evolution of selected global parameters even if these ones parameters are without a realistic physical meaning. However, test is the realisation of a physical phenomenon that may be completely analysed with a suitable interpretation method.

A such aim of work is proposed in the following, in the case of compression and compaction tests. These tests are often used to characterise fine granular media as pastes or powders, in many industrial sectors : food industry, civil engineering, pharmaceutical and cosmetic industries... Results obtained with these easy and rapid tests are nevertheless difficult to understand physically.

The use of the analytical solution known within the simple compression test, with reduced slenderness, on perfect plastic fluid leads to propose an treatment method of the recording data resulting from a test. The goals of the method is to quantify the parameters of a global rheological behaviour of granular media.

Comparison between experimental data obtained with the same sand by the plastometer test (compression test) and the triaxial test validates the proposed method. Interpretation of different compression test results carried out on dry clay powders, from which rheological parameters as cohesion and internal friction coefficient are known, are discussed.

Analogy between simple compression test and pressing test (compaction test) permits the characterisation of powder behaviour tanks to compression test with confinement. Pressing test data on cellulose powders show benefits of the procedure.

2   The plastometer test

Simple compression of a fluid sample, with reduced slenderness, between two coaxial circular and parallel plates, without rotation, is a squeezing flow problem from which the first analytical solutions are given by [1] for Newtonian fluids and by [2] for viscoplastic fluids. In relation with flow of plastic fluid, analytical solutions are proposed by [3], associated with calculus of [4].

For such test and according to [2] and [3] the solution obtained for a Von Mises plastic fluid (plastic yield value K), with sticking condition on wall during the flow, is:

$$F = \pi R^2 K (\alpha + \beta R/h)$$

(1)

where R is the radius of plates, h the height of the sample and F the compression load applied on plates. [2] proposes $\alpha = 0$ and $\beta = 2/3$, for solution corresponding only to the friction effect. $\alpha = 1$ and $\beta = 0$ is the solution of compression without friction on plates. [3] uses $\alpha = 1.5$ and $\beta = 1$. We prefer $\alpha = 1$ and $\beta = 2/3$.

The used solution shows that for Von Mises plastic fluid load F is not influenced by the compression speed v, maintained constant during test. The record of F versus h, R and v are expressed through P versus K. K is calculated with (1) and $P = F/(\pi R^2)$ is the average compression stress. Flow of plastic fluid then appears as a succession of limit equilibriums. The figure 1 shows experimental results. Tests are carried out with ceramic paste [5] with two compression speeds and with different roughnesses of steel plates.

Within the range of compression speed used, the global rheological behaviour of tested paste is essentially plastic. A yield value equal to 120 kPa is identified with scratched plates. Yield value is not influenced by v or P. Use of smooth plates induces a decrease of K, not influenced by variation of v or P. Sliding of paste on the plates, accompanied by a decrease of shearing conditions, accounts for this result. Water content measurement does not show any evolution during compression. Such tests are runing in quite undrained condition.

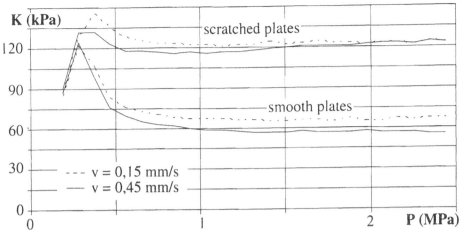

Figure 1: Plastometer tests on ceramic pastes

3 Application on granular media

3.1 Sand tests

Evolution of *P* versus *K*, obtained by a plastometer test on sand gives a first indication to identify the plastic criterion associated to the tested fluid. Such way has been used by [6] to identify the behaviour of mortar (saturated granular media) of which plastic yield value is influenced by effective pressure. Then tests are runing in drained condition.

Tests are made with dry sand (Reguiny sand 0/1 mm [7]). Obtained results are compared with those of a triaxial test, a traditional test of soil mechanics (sand or clay).

Plastometer and triaxial tests are two compression tests where loading of the cylindrical sample is carried up the failure. A judicious use of the compression load record at the time of failure lays out the evaluation of the plastic criterion associated to the tested material. The two tests are controled with a constant compression speed, slow enough to consider the flow of the sample as a succession of limit equilibrium states.

- In the case of triaxial test, the sample is put into a waterproof membrane, saturated and confined (confinement pressure σ_3 is constant during the test). The sample slenderness is large (in the present case height/diameter = 2). Loading is realised with undrained condition and with record of pore pressure. The used compression speed is 0.4 mm/mn.
- In the case of plastometer test, the sample is dry, with reduced slenderness (in present case height/diameter <0.4) and not confined. Used plates are scratched, 50 mm of radius. The compression speed is 0.15 mm/s.

3.1.1   Experimental results

The figure 2 shows curves obtained by the both tests. Results are given in (p',q) coordinates, often used in soil mechanics. The parameter p' is the effective pressure and q is the deviatoric stress.

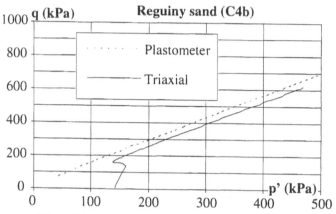

Figure 2: Plastometer test and triaxial test on a fine sand

The curves are linear. Results of both tests lay to associate a same type of plastic criterion to the tested sand: the Mohr Coulomb criterion. Furthermore, parameters values of identified plastic behaviour are similar (cohesion and internal friction coefficient). The calculus method of p' and q from data of measurement, described below, validate the tests comparison.

3.1.2   Interpretation method

In the triaxial test, confinement pressure σ_3, the deviatoric stress q and the interstitial pressure u are recorded. Cylindrical sample is considered as an elementary volume of material loaded by effective principal stresses (global values calculated from σ_3, q and u). The saturation of sample and the behaviour of schearing granular media (dilatancy), induce an evolution of u. Its record permits to follow the succession of limit equilibrium states. The deviatoric stress q is the ratio of the compression load upon the area of the sample section. p' is calculated by :

$$p' = q/3 + (\sigma_3 - u) \tag{2}$$

Exploitation of record obtained during a plastometer test on granular media or powder is broken down into two steps. In the first time, by supposing a plastic behaviour, K is calculated with (1). In the second time, the sample set between two plates is considered as an elementary volume of material loaded by effective normal stresses. The axial stress σ_1 is equal to $F/(\pi R^2)$. The two other stresses are equal to the pressure $K/3^{0.5}$ (necessary to have no confinement). Then, $q = 2K$ and $p' = (\sigma_1 + 2\sigma_3)/3$ are known versus the height of the sample. Identifying a plastic criterion such as Mohr-Coulomb type (cohesion C and internal friction angle ϕ'), parameters of the linear regression of data in (P,K) coordinates are given by:

$$C = K_{(P = 0)}$$
$$(3 - sin\phi') / sin\ \phi' - 2/3^{\,0.5} = \Delta P / \Delta K \qquad (3)$$

With the plastometer test on sand a confinement stress equal to 0 is probably lighly incorrect and induce an under evaluation of p'.

3.2   Tests on kaolin clay powder

Some plastometer test have been carried out on dry kaolin powder with scratched plates and with compression speed $v = 0.017$ mm/s. Some direct shearing tests are realised with the same material. Both show a Mohr-Coulomb criterion with a cohesion $C = 6$ kPa and internal friction angle $\phi' = 43.7°$.

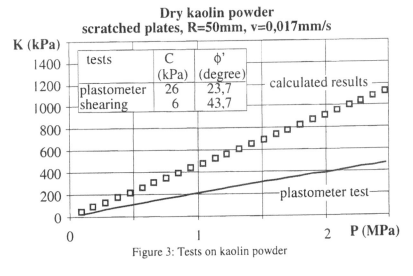

Figure 3: Tests on kaolin powder

The figure 3 permits the comparison between calculated and experimental responses. The two curves are quite linear but parameter values evaluated with (3) are quitly different from those of direct shearing test.

Use of the same interpretation method for the behaviour characterisation of sand or powder seems impossible. However, the observation of the powder sample after plastometer test shows the presence of two distinct zones. A firm central zone of compacted powder is surrounded by a zone of pulverulent powder. The radius of the compacted zone is lower than 20 mm for $R = 50$ mm. Such sample heterogeneity after compression is not observable with sand.

Creation of a firm kernel into the sample requires to reduce the active volume of sample. The average compression stress P must not be calculated with the whole plate surface but with a reduced one. This surface is changing during test. At the begining of the test, the whole sample set between the plates is in the active zone. Sometime after, an active kernel appears. Its radius R', lower than R, decreases as h decreases. By this token, do not take into account the evolution of the active zone induces an under estimation of K for a given value of P. This effect justifies the gap between the two curves on figure 3, for large values of p'.

It is possible to calculate the value of the active zone radius R'. For a given h, R' is calculated laying down P and K to correspond with the plastic behaviour identified by direct shearing test. Calculus realised on data recorded at the end of the test on kaolin powder, with $C = 6$ kPa and $\phi' = 43,7°$ gives $R' = 15$ mm. This calculated value is comparable to the experimental value.

A such interpretation method of plastometer test results is usefull method for the rheological characterisation of powder, provided to take in account notion of active zone in the calculus.

4   Compaction of cellulose powders

4.1   Experimental data

Compaction test is preferred to compression test to do a comparative analysis of powder texture. Compression of a sample set in a rigid cylindrical cell is realised with a ram. Different experimental conditions are usable: constant compression speed, constant rate of load, constant load... As well as plastometer test, recordable parameters are sample height and applied load on the ram.

The use of the interpretation method developed for plastometer test is proposed to analyse compaction test. As an example, experimental data from pressing tests realised by [8] are used. The authors use a pressing apparatus (diameter of ram : 5 mm, maximum height of cell : 25 mm, constant compression speed : 5 mm/s) to test six types of microcristallin cellulose powders (AVICEL PH 101 and PH 102 and VIVACEL PH 101). Such powders are traditional compression binder in pharmaceutic industry. Tests realised by [8] give a graph of the evolution of compression load versus ram moving. Authors consider this type of graph as the most differential.

The figure 4 shows the transcription of experimental results in a graph P versus

H/R if P is the average compression load applied on the ram, H the height of the sample and R the radius of the ram.

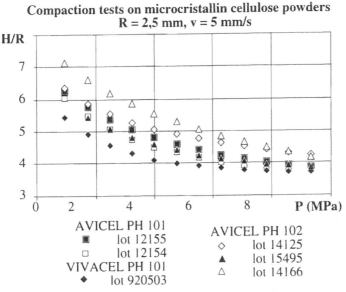

AVICEL PH 101
■ lot 12155
□ lot 12154
VIVACEL PH 101
♦ lot 920503

AVICEL PH 102
◇ lot 14125
▲ lot 15495
△ lot 14166

Figure 4: pressing test on cellulose powders

4.2 Interpretation method

Some hypothesis are proposed:

- the compressed powder cylinder is equivalent to the superposition of n samples subjected to plastometer type tests. Radius of such equivalent samples is equal to the ram radius. Height h of each equivalent sample is linked to H:

$$h = H/n \qquad (4)$$

- the presence of the cell around the powder cylinder induces a confinement effect (radial stress σ_3) proportional to P.
- the plastic criterion associated to the powders is Mohr-Coulomb type. So, structure of (1) is respected with a yield value K function of P and permits to evaluate the applied compression stress σ necessary to induce flow into the equivalent plastometer test:

$$\sigma = (C + aP)(\alpha + \beta R/h) \qquad (5)$$

C is linked to the cohesion and a to the internal friction coefficient.

These hypothesis permit to propose an equation to interpretate the pressing test:

$$AP = (1 + Pa/C)(1 + BR/H) \qquad (6)$$

where A, a/C, B are constant. A is linked to the ratio of the sensibility of material to confinement (connection between axial stress and radial stress) on cohesion. a/C characterises the plastic behaviour. B is directly linked to the number of equivalent samples proposed for the analogy. Use of experimental data validates the interpretation principe and gives the identification of model constants. Assuming $y = 1/P$, the derivative function of (6) gives:

$$z = 1/(\delta y/\delta(H/R)) = (B + H/R)^2/ (AB) \qquad (7)$$

z is calculated for each point of recorded data and $z^{0.5}$ is plotted versus H/R. Results are given in figure 5. The curves are linear and permit to estimate A and B values. These values are negative (table 1). Only a linear regression of the curve of VIVACEL PH 101 sample is debatable. With values of A and B parameters and with (6) it is possible to calculate for each point of recorded data a/C value. Results are plotted versus P on the second graph of the figure 5. Excepted during the begining of the test (low values of P) curves are linear and a/C stays constant. This comment emphasizes the validation of the interpretation method. The VIVACEL PH 101 sample differs from other samples. An explanation is possible considering that this powder has a low cohesion (=>high $|A|$ and high a/C).

Table 1: Values of A, B and a/C parameters

	AVICEL PH 101 lot 12155	AVICEL PH 101 lot 12154	VIVACEL PH 101 lot 920503	AVICEL PH 102 lot 14125	AVICEL PH 102 lot 15495	AVICEL PH 102 lot 14166
A =	-1.67 E-07	-2.43 E-07	-12.20 E-07	-1.03E-07	-2.50 E-07	-1.00 E-07
B =	-7.12	-7.18	-9.58	-6.72	-7.24	-7.64
a/C=	0.95 E-07	1.59 E-07	6.62 E-07	0.71 E-07	1.78 E-07	0.19 E-07

With known parameters of the model it is possible to compare recorded H and calculated H values. The ratio of these two values stays similar to 1 (gap less than 2%).

Taking into account values of model parameters (table 1), (7) is rewritten as follow :

$$(1 + Pa/C)(1 + |B|R/H) = |A|P + 2(1 + Pa/C) \qquad (8)$$

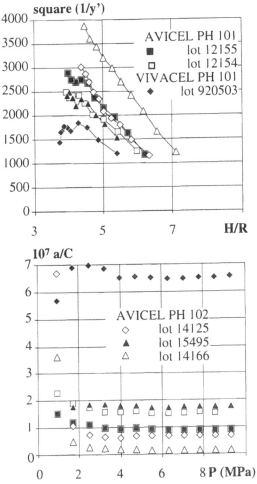

Figure 5: Identification of *A*, *B* and *a/C* parameters

The first part of (8) has structure of the plastic solution of equivalent plastometer test with $|B| = n/\alpha$. The number of equivalent plastometer tests fluctuates between 6 and 1 following the tested powder (with $\alpha = 1$). The second part of (8) combines the plastic behaviour with confinement conditions. The equation (8) links stress state evolution to volume variation of the powder sample.

The used approach permits to modelise the pressing test through three kinds of parameters. It is then possible to understand the physical role of each parameter:

- number *n* is linked to the initial solid volume fraction of powder sample. An initially strongly compacted sample induces a large value of *n*.
- *A* and *a/C*, both influenced by the cohesion, are linked to two further physical principles: the anisotropy of the stress state in the sample, strongly influenced by the powder grain rigidity and the conditions of local grain contacts, and the

pressure condition or drainage condition of the air in the media. Undrained conditions infuenced by granulometry, shape and nature of grains, create an equivalent purely cohesive media.

5   Conclusions

Material behaviour identification is often difficult. However, the use of basic industrial tests can bring a sufficient evaluation. Plastometer test is a good example. Often considered as a comparative test, it is able to become a well adapted identification test in view of plastic materials.

An interpretation method of this test is built within an analytical solution of plastic flow. In the case of firm ceramic paste, sand or powder, the method is successful, provided to take into account a notion of active zone in the calculus.

Considering the analogy between simple compression test and pressing test (compaction test), an adaptation of this interpretation method is proposed. As example, the characterisation of cellulose powders behaviour is studied. Three parameters linked to the solid volume fraction, the anisotropy of the stress state in the sample and the rheological behaviour are calculated. Obtained parameter values give suitable physical indications to compare different tested powders.

References

[1]   SCOTT J.R. (1935): IRI Trans.. 10.

[2]   COVEY G.H. (1977): *Application of the parallel plate plastometer to brown coal rheometry.* thesis. Australie.

[3]   KENDALL K. (1987): *Interparticle Friction in Slurries.* Tribology in particulate technology. edited by Briscoe et Adam: 91-103.

[4]   PASTOR J.. TURGEMAN S. (1982): *Limit analysis in axisymetrical problems numerical determination of complete statical solutions.* Int. J. Mech. Sci.. 24: 95-117.

[5]   MORTREUIL F.X.. LANOS C.. CASANDJIAN C.. LAQUERBE M.. (1997): *Construction et validation d'une méthode de simulation de l'extrusion de pâtes céramiques.* 13ème congrès A.U.M. Poitiers: 11-14.

[6]   LANOS C.. LAQUERBE M.. CASANDJIAN C. (1996): *Rheological behaviour of mortars and concretes: experimental approach.* Production Methods and Workability of Concrete, RILEM: 343-353.

[7]   VUEZ A.. HELLOU M. (1995): *Rapport interne: sable de Reguiny.* GTMa, INSA Rennes.

[8]   CARMONA P.. MICHAUD P.. CORDOLIANI J.F.. CHULIA D.. RODRIGUEZ F. (1996): *Validation d'un analyseur de texture pour la caractérisation rheologique des poudres.* Les Cahiers de Rhéologie, Vol XIV, 4: 807-817.

5 Dynamic experimental investigations

Densification of bulk materials in process engineering

Roman Linemann, Jürgen Runge, Martin Sommerfeld and Udo Weißgüttel

Institute of Process Engineering, University of Halle-Wittenberg, Germany

Abstract: Bulk density and compactibility of bulk materials play an important role in process engineering. Reliable data are required for dimensioning plants to determine volumes of apparatus or for supplying load data for static design. Furthermore, they are used to produce sinter bodies or catalyst beds and to characterise bulk materials. While compression by normal stress can be reproduced and mathematically described, the influence of shocks and vibrations has hardly been investigated. All known standard methods for bulk material compression specify an individual equipment. The parameters of the shocks or vibrations, however, have not yet been defined. So an investigation project was started to examine the influence of defined shocks and vibrations for bulk material compression. Shocks and vibrations can be controlled by an electrodynamic shaker. First results will be presented for powders.

1 State of the art

A charge of particles behaves differently from that of a solid or a liquid. Its main characteristic is that the space which is taken up by it depends strongly on the influence of mechanical forces such as compression, impact and shaking. RUMPF [1] calls this system statuses of packing.

The apparent density of the loosest packing (aerated or loosest bulk density) usually stands for bulk density. SCHWEDES [2] analysed its preparation methods and found that it is complicated to reproduce the adjustment. This state of the loosest packing is observed less often and only for a short time. However, the reproducibility of the loosest packing as an initial state is required to determine compactibility. Therefore the filling procedure of the graduated vessel has to be tested for reproducibility. Selecting the filling method is a matter of experience because the individual properties of the particular bulk materials have to be considered.

In addition to uniaxial compactibility, vibration density and packed or tapped density are used as measures of compactibility. However, precise information to adjust them is not available. Even authors like BUTTERS [3] or WOODCOCK and MASON [4] describe the experimental procedure generally as a measuring cylinder that is exposed to a specific number of taps. They do not even mention vibration densification.

Standards offer information about the procedure (see: [5 to 10]). However, they were usually designed for specific product groups or branches of industry. They have in common that the physical quantities which cause compression are described qualitatively. Consequently, the results of different methods cannot be compared.

The compression of packing by uniaxial pressure can be physically reproduced, and the density curve depending on normal stress can be easily described by approx. 25 different mathematical equations. The equations according to RUNGE and WEISSGUETTEL [11] have a very high fitting quality in the pressure range up to 30 kPa which is relevant to characterise bulk materials.

Some specific publications are concerned with the compression of bulk materials under the influence of different mechanical force actions using CARR´s powder tester [12, 13] as experimental equipment. HARNBY and co-authors [14] measured the influence of the relative humidity of ambient air on loosest bulk density and on the compression ratio according to HAUSNER [15]. Unambiguous dependencies on the humidity of ambient air occurred in the measurements. The authors do not mention in which way they determined the number of shocks. They did not measure the deceleration values. HIROSE and co-workers [16] examined the shock compression of iron powder and found definite dependencies of the density on the drop height and the number of shocks. Like other authors, they do not register deceleration values.

In addition to shock compression, compression by vibration is of great importance in industrial processes. CUMBERLAND and CRAWFORD [17] quote works by several authors who applied different compression methods. The information on frequency and acceleration values is non-uniform and incomplete.

2 Compactibility of expanded Perlit

Expanded Perlit is a vivid example of the influence of particle size distribution and particle form on bulk density and compactibility [18]. It was repeatedly transported by a pneumatic conveyor system and was analysed after one, 20 and 40 conveying cycles. Figure 1 shows the particle size analyses of the products. The degradation of the particle size depending on the number of the conveying cycles can be clearly recognised. Furthermore, the loose bulk density, the vibration density, the tapping density and the uniaxial compactibility of each product were measured. The results are shown in Figure 2.

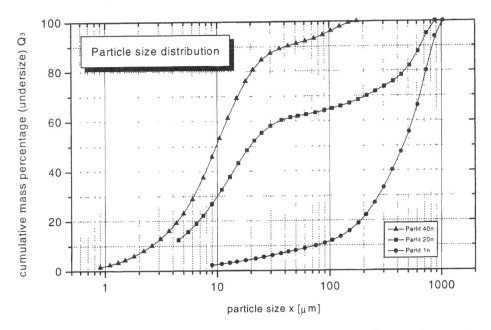

Figure 1: Particle size distribution of Pelit in dependence of the number of pneumatic conveying cycles

With the increasing number of the conveying cycles, the loose bulk density which is the starting point of the compression curves on the ordinate increases. As shown on the lower axis. Compactibility increases depending on the external normal stress. This reflects an enlargement of the rise with an increasing number of the conveying cycles. The curves running approximately linear represent the loose bulk density, the vibration density and the tapping density. They depend on the number of the conveying cycles shown on the upper axis. The vibration density was determined by a sieve machine. The frequency comes to 50 Hertz, the acceleration to 2.5 g and the shaking time to 5 min. The shock table according to DIN 1060 [8] was used to determine the tapping density. 60 shocks were carried out in each experiment. The acceleration amounted to 250 g. The dimensions of the measuring vessel used (95 mm of diameter and 30 mm of height) were those of the JENIKE shear cell.

The results show that the compression behaviour of attrition sensitive bulk materials, such as Perlit, can be judged very well with physically defined parameters of the acting forces. Reliable data for designing apparatuses and for the comparative characterisation of bulk materials are available. Investigations of a larger number of bulk materials [19] showed that it is necessary to intensify research work in the field of compression behaviour. Compactibility measurements with the above mentioned method deepened the understanding of bulk materials but raised even more questions. The most important conclusion is to make the entire and technically reasonable parameter field of the acting mechanical forces (frequency, swinging acceleration and shock intensity) accessible for future investigations.

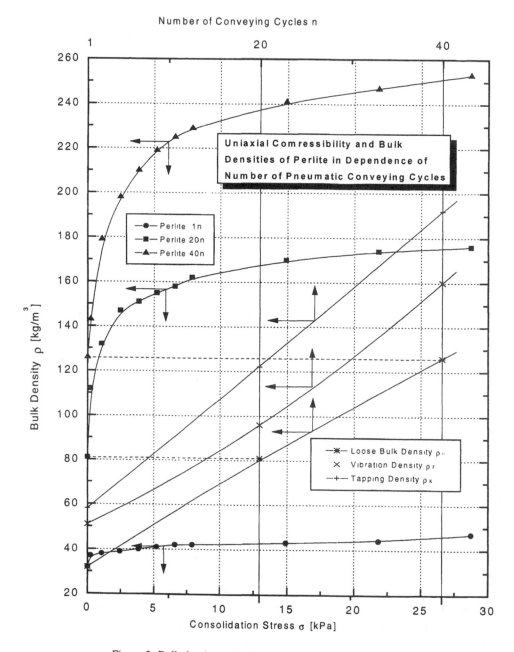

Figure 2: Bulk density and uniaxial compressibility of Perlit

3 Application of an electrodynamic shaker

For the task mentioned above, an electrodynamic shaker is very well suitable. It works on the principle of moving-coil loud-speaker and is able to generate different kinds of vibration and shocks with parameters adjustable in a wide range.

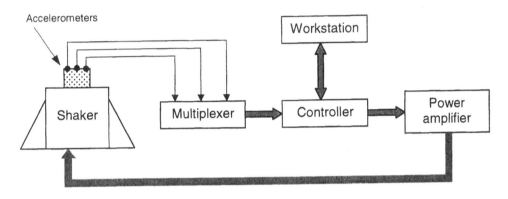

Figure 3: Block diagramm of the vibration test system

Fig. 3 shows a block diagram of the system and Table 1 the performance parameters. The vibration test system is controlled by a workstation. Input and output signals of which are processed by an controller. The power amplifier converts the output signals of the controller into direct current which agitates the magnet coils of the shaker. The vessel filled with bulk material, movements of which can be controlled by up to 4 accelerometers is fixed onto the shakers table. The multiplexer provides the accelerometer signals to the controller which analyses them. Deviations are patched in real time and the parameters are kept within the limits. The process is documented in the workstation.

Table 1: Performance parameters of the vibration test system

Model LDS V830T

Armature diameter	185 mm
Sine force peak	8896 N
Random force	5782 N
Half sine peak bump force	17348 N
Useful frequency range	d.c. to 3500 Hz
Velocity sine peak	2 m/s
Acceleration sine peak	120 **g**
Acceleration random	75 **g**
Max. Acceleration shocks	185 **g**

4 Investigations of highly disperse powders

In the framework of e research program Kaolin powders of a particle range from
2 to 6 μm were analysed. In addition to the uniaxial compression, the powders
were compacted under the influence of vibrations and shocks.

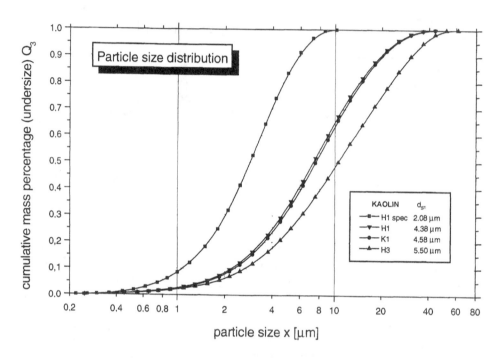

Figure 4: Particle size distribution of Kaolin

The size distribution of the Kaolin powders is shown in Fig. 4. The compression
curves for sinusoidal vibrations (see Fig. 5) show, that the frequency range of the
highest compression can be found at 120 Hz. At higher frequencies density drops.
Further investigations should clarify to what extend it approaches the loose bulk
density. By using the chosen parameters the random mode (Fig. 6) and the uniaxial
compression effect the highest increase in density. High compression rates can be
obtained by shock as well (Fig. 7). The curve sequence of uniaxial compression
(Fig. 8) shows an identical curve form for all Kaolin products. Powders of the
largest SAUTER-diameter (d_{ST}) achieve the highest final density. The example in
Fig. 9 show the compression behaviour of Kaolin H1 spec.powders which was
affected by the three types of vibration tests. Additionally the curve of uniaxial
compression is also presented.

The results show that is necessary to extend the parameter ranges to explain the
compaction behaviour of highly disperse powders. In addition to sine-waves, saw-

tooth and rectangular-wave forms are being planned. Porosity will be applied as a dimensionless measure of compression. Its dependency on the amount of energy involved should be determined and modelled. This requires an intensive occupation with the physical processes which effect powder compaction.

Acknowledgement: These studies were financed by the Deutsche Forschungsgemeinschaft (DFG).

References

[1] RUMPF, H. (1975): *Mechanische Verfahrenstechnik.* München. engl. Edition: *Particle Technology,* Campmann and Hall 1990

[2] SCHWEDES, J. (1968): *Fließverhalten von Schüttgütern in Bunkern,* Verlag Chemie, Weinheim

[3] BUTTERS, G. (Editor): (1982): *Part. Nat. PVC: Form. Structure. Process.* Appl. Sci. Publ., London

[4] WOODKOCK. C. R.. MASON, J. S. (1987): *Bulk Solids Handling.* Leonad Hill. Glasgow u. London

[5] DIN 53466: *Prüfung von Kunststoffen Bestimmung des Füllfaktors der Schütt- und der Stopfdichte von Formmassen*

[6] DIN ISO 3953: *Metallpulver Ermittlung der Klopfdichte*

[7] DIN 25491. *Bestimmung der Schuttdichte und der Vibrationsdichte von Kernbrennstoffpulvern*

[8] DIN 1060. Blatt 3: *Baukalk. Physikalische Prufverfahren*

[9] DIN ISO 3923· *Metallpulver. Teil 1 bis 3. Ermittlung der Fulldichte*

[10] FEM 2 481 *Spezifische Schuttguteigenschaften bei der pneumatischen Förderung.* Definition und Festlegung der Meßverfahren

[11] RUNGE, J., WEISSGUTTEL. U. (1989): *Ein Beitrag zur Beschreibung der Verdichtbarkeit von Schüttgütern bei Normalspannungen bis zu 30 kPa.* Aufbereitungs-Technik: 30: 138-143

[12] CARR. R. L. (1965): *Evaluating flow properties of solids.* Chem. Engineering: January 18: 163-168

[13] MAYERHAUSER. D.: *Der Hosokawa-Pulvertester.* Alpine Aktuell Nr. 57

[14] HARNBY, N.; HAWKINS. A. E.; VANDAME. D. (1987): *The use of bulk density determination as a means of typifying the flow characteristics of loosely compacted powders under conditions of variable humidity.* Chem. Engng. Sci.; 42: 879-888

[15] HAUSNER. H. H. (1967): *Friction conditions in a mass of metal powder.* Int. Journ. of Powder Metallurgy: 4: 7-13

[16] HIROSE. J.; WAKABAYASHI. M.; MURATA, H.(1990): *Effect of wall friction on density distribution in the tapping process.* Proc. sec. World Congress Particle Technology. Kyoto. 19.-20. Sept.. Part 1: 192-198

[17] CUMBERLAND. D. J.; CRAWFORD. R. J (1967): *Handbook of Powder Technology.* Vol. 6 The packing of particles. Ed.: Williams. J. C.; Allen, T.. Elsevier

[18] RUNGE. J. (1993): *Attrition of perlite by pneumatic conveying and its effect on powder properties,* Powder handling & processing 5· 135-138

[19] RUNGE. J. (1993): Ten years of research in the field of mechanics of bulk solids in Merseburg, Powder handling & processing 5: 156-157

Compaction of Loose Flooded Granular Masses Using Air Pulses

Norbert Pralle and Gerd Gudehus

Institute of Soil and Rock Mechanics , University of Karlsruhe, Germany

Abstract: Spontaneous flows of very loose sands pose a serious environmental problem particularly in mining areas. Conventional compaction methods such as blasting densification and vibrofloatation have no or limited monitoring and/or control devices during their application. Alternative methods with monitoring features are needed to remediate those collapsible wastelands in terms of safety. In laboratory experiments the effects of repeated air pulses in loose sands were investigated, focusing on their densification success and liquefaction potential. Tests in a modified semi-large triaxial cell investigated the effect of induced drainage on liquefied sand. Large scale field experiments using an airgun as pulse source were performed on test sites instrumented with geophones and pressure transducers to consider in situ conditions. The results show that repeated air pulses can liquefy and then compact cohesionless soils. The extent of liquefaction can be monitored on-line and controlled through induced drainage. The proposed method may be useful when careful soil treatment is required because of closeby sensitive structures.

1 Introduction

Strip mining in former East Germany has created large wastelands of artificial soil deposits of highly loose material. Closing of the mining activities led to termination of the ground water drawdown. As the ground water level rises the stability of the sands is further lowered. Large holes owing to the coal mass deficit are being filled with ground water as it rises forming large lakes. Their shore lines are in a particularly critical state despite their gentle slope of a few degrees. CPT investigations along the shore lines and the hinterland confirmed very loose sand with typical tip resitances as low as 2 MPa [5]. Sampling using the soil freezing method revealed significant spatial density fluctuations [9] that contribute strongly to the collability of loose sands [8]. For major parts of the deposits, which can reach maximum depths between 60 and 70 m [4, 6] an economical and efficient method to stabilize the ground is being conducted using the blasting method. However, the gently sloping shore lines are highly collapsible and thus are very prone to settlement liquefaction. The stabilization of these areas turned out to be a difficult task, since any densification procedure implies disturbances that can cause further liquefaction flows.

Any densification measure of loose sand requires a temporary liquefaction of the soil , i.e. for a certain time the soil is further destabilized. To encounter this difficulty vibrofloatation and stepwise blasting measures were conducted to carefully densify in order to remediate the area. Blasting densification bears a high degree of uncontrollability, since the explosive charges have to be designed beforehand. Adjustments during the procedure are thus not possible. Vibrofloation is a much gentler procedure since the generated liquefaction is local and can be stopped at any time. However, this procedure requires constant supply of sand to refill the generated voids. Both densification methods lack a control tool. Hence, an effective and safe method should include an optional control.

Densifying sand using repeated air pulses is not a particularly new method. Trichonow [15] densified saturated sands using alternating high pressured air pulses that were introduced through steel tubes into the ground. Stoll et al. [12] suggested a densification method using an airgun as pulse source in combination with horizontal drilling in order to stabilize loose sands from a safe distance. However, these methods lack a method to adjust for the spatial variation of the sand's collability.

This study reports on the effects produced by high pressure air pulses on sand. The study focused on local liquefaction and densification of loose sands caused by single and sequential air pulses in partly saturated sands. The first part describes laboratory experiments under controlled boundary conditions, such as density, pressure, granulometric properties of the material. With a pneumatic dynamic probe (PDP) air pulses were released at designated depths. The second part describes two field tests conducted in loose sands of the mining deposit areas where air pulses were introduced into the ground down to maximal depths of 17 m using an airgun. The test sites were instrumented with 3-D geophones, pore pressure as well as earth pressure transducers, yielding time series of particle velocities, pore water pressures, and earth pressures.

The release of short repeated pulses in saturated or partly saturated non-cohesive soils causes an increase of the pore water pressure within a contained suspension bubble. Its extent depends on the ambient pressure, the pressure level and the number of pulses. The duration of the liquefied zone depends on the potential for drainage. During the diffusion of the pore water excess pressure the grain skeleton is being reorganised at a higher density, leading to a higher stability. Since drainability of the soil is part of the densification process careful monitoring of that aspect is suggested as a control tool. Hence, artificially induced drainage can serve as a limited tool to control the potential collability of the soil.

2   Laboratory Experiments

Two lab experiments deal with the assessment of densification and the extent of liquefaction due to air pulses. A third experiment deals with the controlability of artificially liquefied sand samples at critical state through induced drainage.

2.1 Local densification through repeated air pulses

This experiment aimed at the local densification of sand due to repeated air pulses. In a large calibration chamber filled with very loose, partly saturated sand air pulses were introduced. Density changes were verified with CPT and the evaluation of settlement [14].

2.1.1 Method & Materials

Fig. 1 shows a barrel-like calibration chamber of wooden beams and steel rings providing support for a large rubber membrane that kept the system water tight. Its base provided for water supply and drainage devices necessary for flooding and draining the sample.

Figure 1: Large calibration chamber for air pulse experiments. Height 130 cm, ⌀ 100 cm

A pneumatic dynamic probe (PDP) served as air pulse source (Fig. 2). The PDP consisted of a metal tube with a standardized CPT tip. The tube volume served as pressure chamber for compressed air. Using a dead load the PDP penetrated the soil to a designated depth. After releasing air pulses by triggering a mechanical valve, the tip resistance q_c was measured at serveral points.

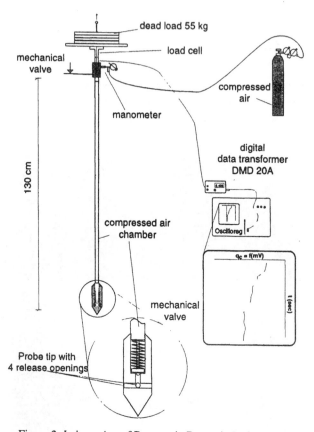

Figure 2: Lab version of Pneumatic Dynamic Probe (PDP)

Six experiments (V1 - V6) were performed all with a sample 100 cm in height and 100 cm in diameter. With a large funnel and careful procedure the chamber was filled with Stuttgart Sand (e_{max} = 0.99; e_{min} = 0.57) obtaining densities as low as e = 0.98. The remaining cone was flattened using vaccuum suction. Slow flooding up to 10 cm below its surface produced no observable sagging of the sample and degrees of saturation S_r between 0.89 and 0.98. Before releasing the air pulses a CPT was performed in the undisturbed sample in its center as reference. Several air pulse series were then released, CPT and settlement measurements conducted after that according to the sounding points shown in Fig. 3.

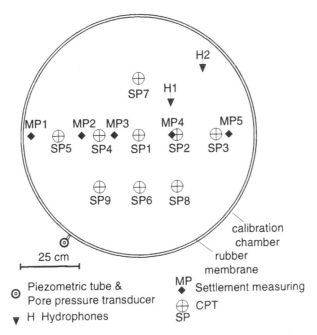

Figure 3: Map view of calibration chamber filled with sand. Shown are points for settlement measurements and CPT.

On-line control during the experiment (time series) was provided for settlement measurement (at one point), global and local pore water excess pressure (using sensitive hydrophones). As for global pore water pressure we refer to the overall average pressure of the whole sample.

The experimental programme was performed as follows:

- CPT at point SP1 in undisturbed soil for a reference value.

- Release of 3 air pulses within 5 seconds. In most experiments the air pulses were released in a depth of 60 cm.

- Settlement reading at points MP to determine crater size

- On-line measurement of global and local pore water pressure

- CPT at points SP after the air pulses.

- Soil sampling according to DIN 18126 (V5) and using the soil freezing method (V6)

2.1.2   Results & Analyses

Immediately after releasing the air pulses the global pore water pressure increased, which could be observed through the piezometric tube. Concentric cracks around the PDP developed at the surface, and were shortly after covered with sand boils (Fig. 4). The total liquefaction of the immediate PDP tip environment caused a sudden sagging of the probe. The water level rose and covered the whole sample.

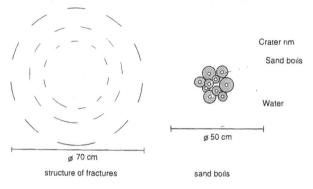

Figure 4: Air pulses producing concentric cracks and sand boils

Fig. 5 shows the settlement trough of experiment V2. Flooding of the sample caused a negligeable settlement. However, it can be clearly seen that each impulse series led to major settlements. This data served as base to calculate the global void ratio changes.

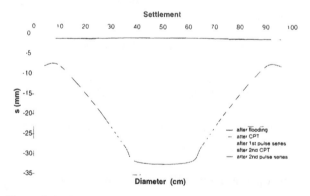

Figure 5: Settlement profile after introducing 2 series of air pulses

The spatial density variation was determined using CPT at various points (Fig. 3). However, when performing CPT in a calibration chamber the ratio between tip diameter d_{tip} and chamber diameter D_{cc} has to be considered. The denser the sample the larger has to be D_{cc} in order minimize the influence of the confining chamber. If D_{cc}

is too small q_c will be too low compared as if no rigid boundary was present [7]. For very loose sands D_{cc} should be 20-40 times d_{tip}. This value was observed for these experiments. However, with decreasing distance to the chamber walls q_c is slightly exaggerated since the sand can not escape the intruding tip into all directions. In addition, CPT itself influences the state of the soil. Loose sand is densified, dense sand will be loosened. Which influence the CPT had on the following soundings was not considered.

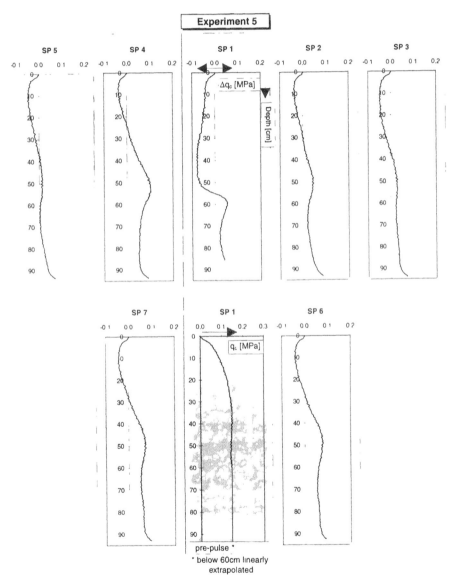

Figure 6: Tip resistance difference Δq_c for all CPT sounding points SP (Exp. #V5)

Fig. 6 shows the tip resistance differences Δq_c for experiment #V5, i.e. from the post-pulse CPT-profiles the CPT-value prior to the pulse disturbance at point SP1 was subtracted. Negative Δq_c stands for loosening, positive for densification. At point SP1 Δq_c is negative down to the initial depth of the tip due to the previous retrieval of the PDP. However, the densification effect of the air pulses can be clearly seen, particularly around the pulse source area, where locally max. values of 80% densification increase were obtained (see SP 4, at 55 cm depth). Furthermore, a general loosening can be observed in the approx. upper 25 cm of the sample. This can partly be attributed to the risen water level (the reference value was obtained in a sample with the water level being 10 cm below sample top) and the sand boils that contribute to the loosening of the sand.

2.2   Liquefaction as precursor of densification

While the previous experiment showed densification to be a result of the pulse-induced liquefaction it was now of interest to investigate the spatial extent of liquefaction and to "look" into a liquefaction zone [3]. Dynamically induced liquefaction of sands for example due to earthquakes has been broadly investigated [2], [1]. Studer et al. [11] , however, investigated blast-induced pore water excess pressures and its resulting liquefaction. Introducing a semi-dynamic liquefaction coefficient or pore water pressure ratio (PPR) yields a relation between the artificially induced excess pressure u_{ex} and the ambient effective stress σ_0'. With PPR $= u_{ex} / \sigma_0' = 1$ the effective stresses vanish, i.e. the soil turns into a suspension.

2.2.1   Method & Materials

The set-up was similar to the previous experiment. High-sensitive hydrophones were modified to dynamically measure local relative pore water and total excess pressure changes in the sand that was being subjected to air pulses through the PDP at a designated depth. After the measurements samples were taken with the soil freezing method in order to obtain an insight into the topology of a liquefaction zone. For better illustration the sand was prepared with marker horizons.

The experimental programme was performed as follows:

- Preparation of loose sand with marker horizons. Determination of void ratio e and S_r before air pulse release.

- Installation of hydrophones at same depth as air pulse source serving as local relative pore water pressure and total pressure transducers.

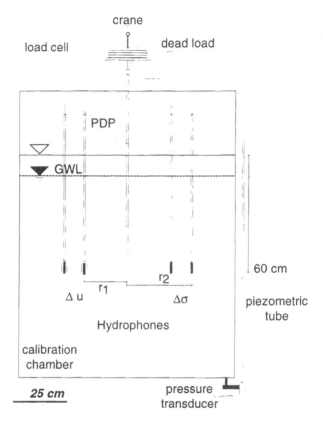

Figure 7: Set up and instrumentation to investigate the extent of liquefaction zone

- Installation of an absolute pressure transducer for global pore water pressure measurement.

- One pre-pulse CPT profile in center of sample for comparison purposes.

- Release of 3 air pulses within 5 seconds in 60 cm depth.

- On-line data acquisition of local and global pore excess pressure changes as well as local total stress changes.

- Sampling using the soil freezing method.

2.2.2 Results & Analyses

Fig. 8 shows the development of global and local pore water excess pressures during the air pulse disturbance. Locally the pore water excess pressure increases immediately, whereas the global pore water increase is a function of the sand's permeability. The global u-curve shows weak reduction peaks in the very moment of the pulses, which are likely to be due to a short pulse induced volume increase. Noticeable is the further pore pressure increase by ca. 100% after the last pulse.

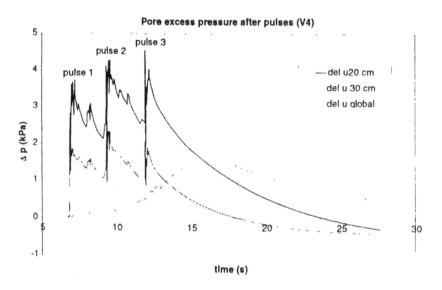

Figure 8: Global and local pore water excess pressure vs. time (Exp. #V4)

The hydrophones record pressure changes. Since the process of grain reorganisation is rather slow, total stress changes can be attributed to wave propagation and pore water excess pressure increase. Fig. 9 shows PPR-values being larger or equal 1 up to a radius of 20 cm from the pulse source. At a distance of radius of 40 cm effective stresses were reduced by 50%.

Four samples of 100 cm length were retrieved for each experiment using the soil freezing method. The samples were cut open and owing to the marker horizons the liquefaction zone could be measured. Surprising was the fact that even in the center of the liquefaction area the marker stratification was preserved (Fig. 10). The pore water excess pressure destroys the grain contacts producing a suspension consisting of sand, water and gas without changing the original topology.

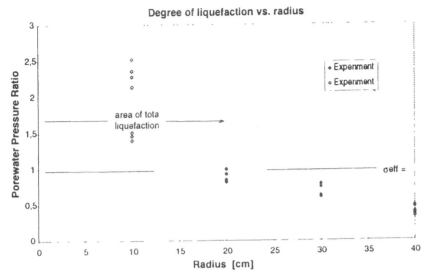

Figure 9: Pore water pressure ratio vs. distance (Exp. #V3,4)

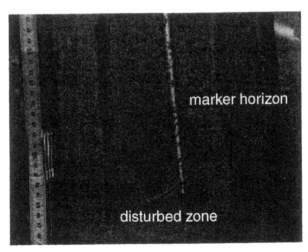

Figure 10: Frozen open cut sample with horizon markers showing deformation topology of liquefaction zone

2.3   Controlling artificially induced liquefaction

If liquefaction is the necessary precursor for densification and could artificially be stimulated then artificial drainage could possibly serve as a tool to control or possibly inhibit a catastrophic extent of liquefaction. The goal of this lab experiment was two-fold: The collability of soil at critical state when subjected to air pulses and the effect of induced drainage after partially liquefying a sand sample [10].

2.3.1   Method & Materials

Semi-large triaxial sand samples ($\oslash$ ca. 20 cm, height ca. 40 cm) were instrumented with relative pore water and earth pressure transducers (modified hydrophones). The samples were subjected to a dead load simulated with using a pneumatic prop attached to a steel frame (Fig. 11). Dead load and cell pressure were adjusted to set the sample into a critical state (K_0-state).

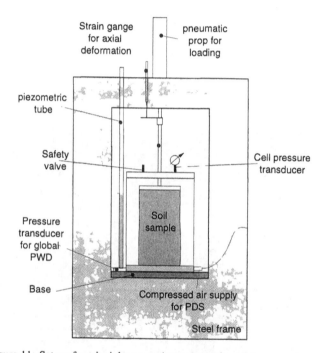

Figure 11: Set-up for triaxial tests on instrumented semi-large sand samples

Instead of conventional strain or force controlled cyclic tests the samples were dynamically perturbated introducing air pulses into the sample. For that purpose a small PDP was installed inside the sample; the air pulses were triggered mechanically. On two levels pore water excess pressure and total pressure changes were measured with hydrophones (Fig. 12). Furthermore, absolute global pore water pressure was recorded as well as the axial deformation and its rate. Seven experiments were conducted with similar void ratios and degrees of saturation, but different drainage conditions (see table 1).

Table 1: Experimental parameters

Exp. #	e [-]	S_r [-]	Drainage
2	0,877	0,75	open
3	0,89	0,75	open
4	0,9	0,81	open
5	0,88	0,81	closed/open
6	0,87	0,81	closed
7	0,88	0,81	closed
8	0,87	0,81	open

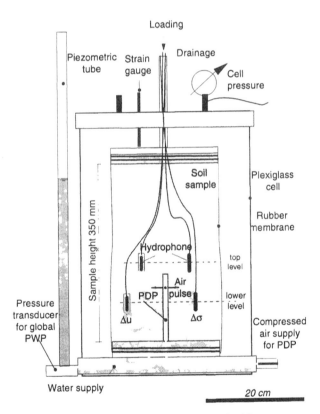

Figure 12: Cross section through triaxial sample, instrumented with pore water and total pressure transducers on two levels. The optional drainage was realized through vertical loading axis.

2.3.2   Results & Analyses

Figs.13 and 14 show time series of effective stresses and the cumulative deformation for two samples subjected to dynamic disturbances without and with drainage, respectively. The cumulative vertical deformation for the sample with drainage (Fig. 14) is much smaller (ca. 70% less) than for the testing sample without drainage (Fig. 13).

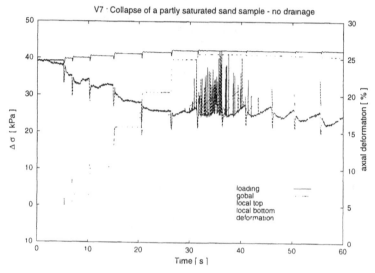

Figure 13: Effective stresses and vertical deformation during air pulse induced perturbation (no drainage).The erratic data between 30 and 42 s were due to an momentary short circuit.

However, it is noticeable that there is no difference for the deformation rate during the first three air pulses. Hence, drainage requires a start-up time. Furthermore, it is shown that liquefaction is also possible for partially saturated samples (see table 1); experiments with completely saturated samples were not performed though.

It is shown that sand subjected to dynamic disturbances has a much lower collability if a possibility to drain is given. Furthermore, it has been shown that repeated disturbances can "pump up" pore (water) pressure to such extent of sample failure. Hence, it may not be necessary to increase pore water pressure with one single blast, but it is possible to compact using repeatedly "gentle" pulses.

3   Field Experiments

The results of laboratory experiments are difficult to transfer to in situ conditions. Scaling laws are almost impossible to apply for dynamic processes. Hence, two

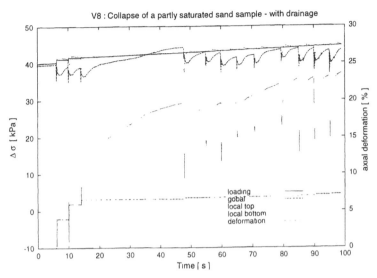

Figure 14: Time series of effective stresses and vertical deformation during air pulse induced perturbation - with drainage

field experiments were conducted on test sites located on deposits of loose sands in order to investigate the effects of high energy air pulses in an in situ environment and to possibly draw correlations to the lab test results. Test #1 dealt simply with the feasibility of introducing high pressure air pulses into the ground using a modified commercial airgun. The test sites were instrumented with 3D geophones, pore water and earth pressure transducers. According to the lab tests air pulses at different pressures and frequencies were released in the ground at different depths. Particle velocities, pore water and total pressures were recorded and analysed.

3.1 Pneumatic Dynamic Probe (PDP) - field version

In order to scale up the laboratory experiments testing equipment as well as the set-up had to be completely redesigned. Fig. 15 shows schematically the main features of the PDP field version.

The tip consisted of a casing protecting a commercial airgun[1] and was attached to a steel tube of 17 m in length hosting the air and electrical supply hose and cables, respectively (Fig. 16). Standard operation pressure of this airgun was approx. 200 bar

---

[1] Airgun VLAS - M1.0.0 by Prakla Seismos

with an average pulse duration of 2 ms. For further details on the working principle of an airgun see Telford et al. [13]. For test #2 earth and pore water pressure transducers as well as an optional drainage were added to the PDP body (Fig. 15).

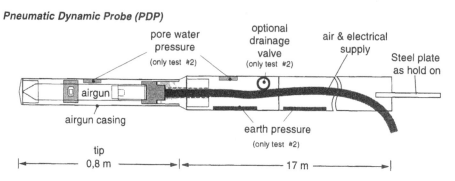

Figure 15: Main features of the PDP field version

Figure 16: Airgun mounted at the tip of the PDP (casing lying beneath)

Because of the large tube diameter ($\oslash$ 180 mm) the PDP could only be vibrated into the soil. In order to operate and manoeuver the PDP (total weight including vibrator engine amounted to 7 tons) a heavy truck crane was necessary.

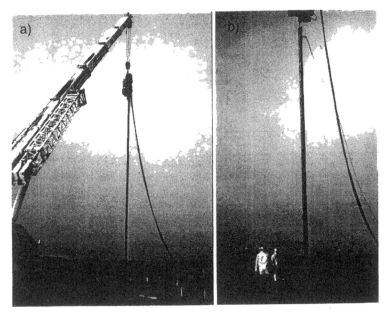

Figure 17: Heavy duty truck mounted crane manoeuvring the PDP during the experiment. Operation radius approx. 20 m

Fig.17 shows a truck mounted crane operating the PDP (a). Fig.17b shows the PDP shortly before penetrating the soil. At the top the vibrator is shown.

3.2   Test sites & operation

Both test sites required a stabilized base for the PDP operating crane. Test #1 was conducted towards the hinterland, test #2 along the sloped shore line at a different location (Fig. 18). The test sites were instrumented with pore water transducers and 3D-geophones. In test site 1 the instruments were deployed in two levels (11 m and 15 m below ground level (GL)), in test site 2 only in one level (15 below GL), and transducers for measuring total stresses added. On both sites surface geophone stations were deployed.

The PDP was vibrated into the ground according to the map on Fig. 18 down to the deployment depth of the instruments. Having reached the given depth air pulses were released with different pressure levels and frequencies. Pulse pressures and frequencies were varied. For each experimental unit particle velocities at all geophone stations as well as pore water pressures were recorded. Since test site #2 was

Cross section

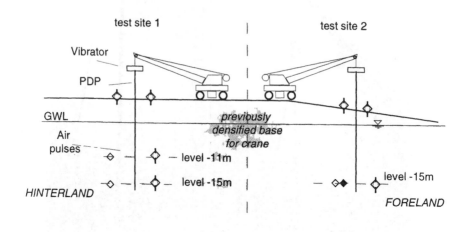

Map View

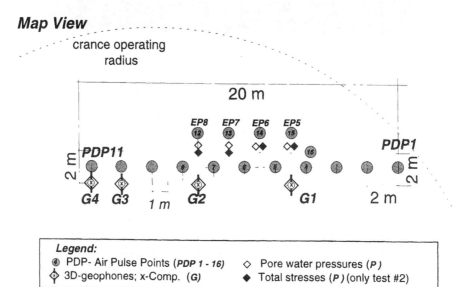

Figure 18: Schematic cross section and instrumentation map of test sites. Note, the tests were performed at different locations

additionally supplied with earth pressure transducers time series of effective stresses could be directly obtained.

3.3   Results

Preliminary CPT and pressuremeter investigations revealed very low densities for both sites. Typically, tip resistances of 2 MPa were encountered. However, the soil of test #2 contained a few percent of fines (ca. 7%), which had a significant impact on the dynamic behaviour of the soil. The results are presented in following categories:

3.3.1   Observations

The compressed air supply turned out to be insufficiently powerful for the given purpose. Fig. 19 shows the development of the supply pressure starting at a pulse pressure of approx. 190 bar and declining by ca. 15% during this experimental unit.

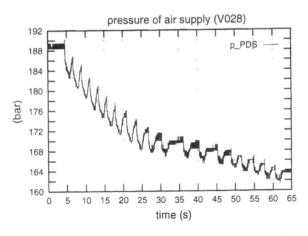

Figure 19: Airgun supply pressure (Test site #1 - Exp.V028)

Nevertheless, during the tests on site #1 the pulse induced ground motions seemed to increase with every pulse. After 4 or 5 pulses the ground started to shake as if made of jelly at a distance of ca. 30 m from the pulse source. This subjective perception was confirmed by the particle velocity recordings. Further pulses increased the pore water pressure to such degree that water suddenly gushed through the top of the PDP steel tube. After several hours a settlement trough began to develop.

3.3.2 Particle velocities

Fig. 20 shows typical time series of particle velocities v_p recorded at both test sites at the same depth (15 m below GL), with same pressure level and frequency as well as similar disturbance duration and distance to the pulse source.

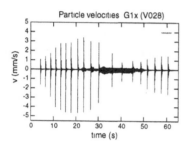
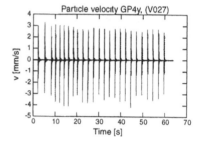

(a) Site #1, V028, G1x, r=11.5 m   (b) Site #2, V027, G4y, r=12 m

Figure 20: Time series of particle velocities v_p due to air pulses from both test sites (r distance to pulse source, x,y are horizontal components, recording depth was 15 m below GL)

The recording from test site #1 (a) shows a "fish-bone" characteristics, which cannot be observed at test site #2 (b). The particle velocity v_p decline after approx. 25 seconds (a) correlates well with the drainage inception. However, the soil of test site #2 has a lower permeability and neither drainage nor a v_p increase could be observed (Fig. 20b). Fig. 21 illustrates well the behaviour of the velocity maxima v_{p-max} for all 3 components of geophone 1 and the pressure during the experiment V028 on test site #1. The peaks increase by approx. 400% of their initial value despite of the source energy decrease. Striking is the fact that all components show the same qualitative behaviour.

Fig. 22 shows the v_p time series recorded during the same experiments, but with surface geophones stations. The same v_p-peak increase can be observed (the strong high frequency content between 22 and 50 seconds was due to an intermediate pulling out of the PDP to a higher level using the vibrator. Being closer to GL explains the higher v_p that follow). The particle velocity increase was generally observed, regardless of the recordings station distance from the pulse source.

3.3.3 Pore water and earth pressures

Just before the air pulses were triggered the pore water and earth pressure transducers recorded correctly the ambient pressure level according to their position in

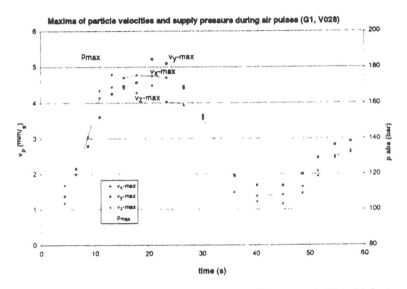

Figure 21: Maxima of particle velocities for exp.#028 on test site#1 (r=11.5 m)

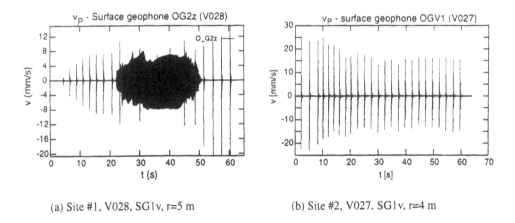

(a) Site #1, V028, SG1v, r=5 m (b) Site #2, V027, SG1v, r=4 m

Figure 22: v_p recorded with surface geophones

the ground. Fig. 23 shows the water pressures for exp. V028 on test site #1 and exp. V027 on test site #2. Before the pulse induced pore water excess pressure started to rise in both cases the ground water level was recorded. Then both pore water pressure time series show an increase of approx. the same magnitude, whereas the pore water excess pressure at site #2 rises rather slow compared to the one on site #1.

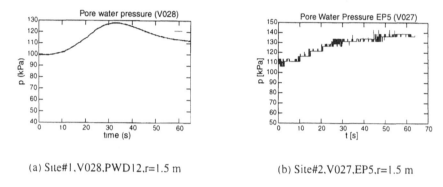

(a) Site#1,V028,PWD12,r=1.5 m   (b) Site#2,V027,EP5,r=1.5 m

Figure 23: Pore water pressure development at a distance of 1.5 m from the pulse source - Site #1 (a), Site #2 (b)

3.4 Analysis

The field experiments confirm qualitatively the findings of obtained in the lab. Non-cohesive soil can be liquefied and thus densified using air pulses. However, if even small amounts of fines are present the liquefaction potential is strongly reduced or even hindered. The field experiments can be summarized with Fig. 24. The main instrumentation is located at the same depth as the pulse source. Repeated air pulses produce a suspension bubble (a). Fig. 24b depicts the particle velocity increase as a function of the extent of the suspension bubble. As the bubble grows the v_p attenuation at a certain spatial point is reduced. The first pulse only liquefies the immediate area around the source, thus improving coupling between soil and source. Part (c) shows the time series for the pore water excess pressure at three distances.

Since the air pulse energy is constant (actually diminished slightly) the behaviour of v_p must be a function of the suspension bubble. The vertical component of the surface geophones show qualitative the same behaviour, regardless of their distance to the pulse source.

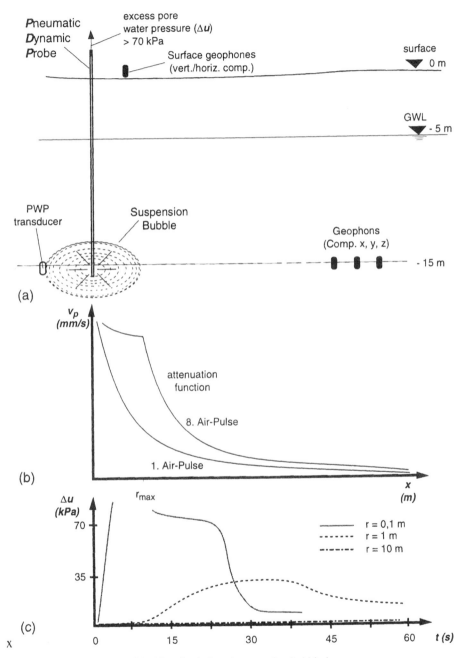

Figure 24: Air pulse induced suspension bubble increase

4 Discussion & Conclusion

Laboratory and large scale field experiments show that repeated air pulses can produce a controlled liquefaction zone in loose non-cohesive, partly saturated sands, which after reorganisation will yield a higher density than before liquefaction, however, without any difference in topology. Furthermore, induced drainage served as a control tool to reduce deformation once the soil was liquefied and at its failure limit. Field tests confirmed these findings using particle velocity recordings. Under the condition of the soil being dynamically liquefiable, i.e. cohesionless and without fines an increase of particle velocity induced through dynamic disturbances could be obverved. As soon the growing suspension bubble found a draining escape, the suspension bubble collapsed and the particle velocities decreased. Tests in soil with lower permeability showed little v_p increase and almost no effect due to drainage. That could mean that the suspension bubble does exist, however with very little extent. Since this behaviour is also observed at surface geophones monitoring by using particle velocities can be conceived when working in a dangerous or sensitive area. The tests show furthermore that small amounts of fines reduce the liquefaction potential strongly. That is to say, that it may take more time to liquefy soil of that kind, but also will be more difficult to control using drainage.

References

[1] ARMIJO G., SOLA P., SERRANO C., OTEO C. (1994): *Review of methods for evaluating liquefaction potential based on the observed behaviour in previous earthquakes.* Earthquake Resistant Construction & Design, Savidis (ed.) 1994 Balkema. ISBN 90 5410 392 2. 14

[2] ISHIHARA K. (1993): *Liquefaction and flow failure during earthquakes.* Géotechnique, Vol. 43, No. 3, pp. 351-415. 14

[3] JETSCHMANEGG I.(1997): *Laborversuch zur Bestimmung des Verflüssigungsbereiches nach Eintrag eines Luftimpulses in locker gelagertem, teilgesättigtem Sand.* Diploma Thesis, Institut für Boden- und Felsmechanik. Universität Karlsruhe.

[4] LEONHARDT A., FÖRSTER W. (1993): *Das Verflüssigungsverhalten von Sand unter Einwirkung dynamischer Lasten.* VEB, Deutscher Verlag für Grundstoffindustrie. Stuttgart S. 63-135.

[5] LMBV (1997): *Remediation an securing slopes and sand deposits prone to settlement flow.* Internal Report.

[6] NOVY A., REICHEL G., WARMBOLD U., VOGT A. (1999): *Geotechnische Untersuchungen und Verfahren bei der Sicherung setzungsfließgefährdeter Tagebaukippen der Niederlausitz.* Braunkohle, Surface Mining, Vol. 51, No. 4, Jul/Aug 1999.

[7] PARKIN A. AND LUNNE T. (1982): *Boundary effects in the laboratory calibration of a cone penetrometer in sand* Proc. 2nd Europ. Symposium on penetration testing, Amsterdam, Vol.2, pp 761-768.

[8] PRALLE N. (1999): *Void ration fluctuations in loose sands - responsible for significant loss of shear strength?* Proc. 13th. Int. Young Geotechnical Engineers Conference, Santorini, Sept. 1999.

[9] PRALLE N., BAHNER M.L. & BENKLER. J. *Computer Tomographic Analysis of Undisturbed Samples of Loose Sands Reveal Large Gas Inclusions in Phreatic Zone.* In preparation.

[10] SANWALD M.(1998): *Laborversuche zur Untersuchung der Stabilität einer locker gelagerten Sandprobe bei dynamischer Belastung.* Diploma Thesis, Institut für Boden- und Felsmechanik, Universität Karlsruhe.

[11] STUDER J., KOK L. (1980): *Blast-induced excess porewater pressure and liquefaction. Experience and application.* International Symposium on Soils under Cyclic and Transient Loading/ Swansea/ 7.-11. Jan. 1980.

[12] STOLL R.D., TUDESHKI H.H. (1996): *Möglichkeiten des Einsatzes der gesteuerten Horizontalbohrtechnik bei der Sicherung von setzungsfließgefährdeten Tagebaukippen.* Vortrag auf dem 6. Merseburger-Umwelt-Tag 1996, Berichte der Fördergemeinschaft Ökologische Stoffverwertung e.V. Halle: Berichte 1/96. 11

[13] TELFORD W.M., GELDART L.P., SHERIFF R.E. AND JKEY D.A. (1988): *Applied Geophysics* Cambridge University Press.

[14] THEILE T.(1996): *Laborversuche zur Analyse des Verdichtungseffektes durch Eintragung von Luftimpulsen in lockerem, teilgesättigtem Sand.* Diploma Thesis, Institut für Boden- und Felsmechanik, Universität Karlsruhe.

[15] TRICHONOW-JAKOLEW D.A. (1959): *Densification of saturated sands with a method using air pulses* Leningrad (Russian).

6 Theoretical approaches

Compaction dynamics of tapped granular matter

A mean field model and its Langevin extension

Annekathrin Döhle and Stefan J. Linz

University of Augsburg, Germany

Abstract: Granular systems can compact under the impact of sufficiently strong, successive, periodic tapping. Recent experiments have revealed that the (ensemble-averaged) packing fraction obeys a very slow, inverse logarithmic relaxation with time or tap number to a final, dense packing fraction. We provide the corresponding macromechanical relaxational law that is based on simple macromechanical arguments. We also discuss a stochastic extension of this relaxation law mimicking the dynamics of individual compaction processes.

1 Introduction

Granular matter such as dry sand or powder consists of large collections of dry massive macroscopically extended particles that basically interact only via repulsive forces. Due to the absence of significant attractive forces between the grains, large enough external forces such as tapping or shaking can lead to a loosening and a reorganization of the grains. As a result of such forcing, a densification of granular matter by reducing the void volume between the grains, commonly called compaction, can occur and shows very evidently the distinction of granular matter from ordinary solids and Newtonian fluids.

Despite its obvious technological and practical importance (cf. [1]), detailed quantitative experimental studies on the compaction behavior of granular matter under tapping have started only recently. The group of Jaeger and Nagel [2] have investigated the settling of monodisperse glass particles in a vertical tube into more compacted states by applying periodic vertical tapping with a tapping intensity $\Gamma > 1$ which is given by the ratio of the peak intensity of the tap and the gravitational acceleration. In these experiments [2], the data for the increase of the packing fraction ρ (being the ratio of the grain space and the total volume of the granular system) with time or tap number t (averaged over several individual compaction processes) could be fitted to the functional form

$$\rho(t) = \rho_\infty - \frac{\rho_\infty - \rho_0}{1 + B \ln(1 + t/\tau)}. \tag{1}$$

Here, B and τ are coefficients that depend on the tapping intensity Γ, but not on the time t [2]; $\rho_0 = \rho(t = 0)$ and $\rho_\infty = \rho(t = \infty)$ denote the packing fraction of the initial loosely packed state and the final packing fraction, respectively. In a subsequent study [3], also the statistical properties of individual compaction processes have been experimentally analyzed. In this contribution, (i) we present an elementary relaxation model in form of a first order nonlinear differential equation for packing fraction that explains – on a macromechanical basis – the result (1) and (ii) we give some general insights in the modeling of individual compaction processes by using a stochastic differential equation approach.

2   Minimal Relaxational Law

In order to state the minimal relaxation law for compaction of periodically tapped granular matter, it will be convenient to use the *reduced packing fraction*

$$A(t) = \frac{\rho(t) - \rho_\infty}{\rho_0 - \rho_\infty},$$ (2)

being a quantity that varies between $A(0) = 1$ and $A(\infty) = 0$ during the compaction process. The substantiation of the relaxation law for $A(t)$ leading to the inverse logarithmic relaxation of the packing fraction (1) [2] is based on the combination of *two* separate microscopic mechanisms being dominant on different time scales: (i) biased void diffusion [4] and (ii) collective reorganization [5, 6]. For alternative motivations of the relaxation law, we refer to [7].

Biased void diffusion or the annihilation of void space by destruction of arches and bridges in the granular system has been modeled in [4] using a one-dimensional nonlinear diffusion equation for the void volume. In this work, it has been found that the settling of the height $z(t)$ of a grain-filled column decays linearly and within finite time or tap number t_c to a final compacted state $z_c = z(t_c)$, $z(t) = z(0)[1 - pt]$ for $0 \le t \le t_c$. Using the fact that the packing fraction $\rho(t)$ is related to the height $z(t)$ by $z(t) = V_g/S\rho(t)$ with V_g being the grain space and S the cross section of the tube, the linear decrease of $z(t)$ transfers to an algebraic increase of $\rho(t)$, $\rho(t) \propto 1/[1 - pt]$ until t_c has been reached. Obviously, biased void diffusion cannot reproduce the correct long-time behavior of (1). Its short-time behavior, however, coincides with the experimental fit (1). Rephrased in terms of the reduced packing fraction $A(t)$, this suggests that the biased void diffusion process is determined by a relaxation law $\dot{A} \propto -A^2$.

Collective reorganization [5, 6] is based on the idea that an increase of the packing fraction of an already densely packed system requires a collective rearrangement of a large part of the grains in the system in order to achieve a more compact arrangement. It is reasonable to suppose that this effect is exponentially costly for particles that can move independently and randomly. Rephrasing this argument [5, 6] in terms of the reduced packing fraction yields a decay behavior $\dot{A} \propto \exp[-(1 - A)/kA]$ with k being a coefficient. Although this equation reproduces the long time decay of (1),

$A(t) \propto 1/\ln t$, it does not yield the full functional form (1) for the whole compaction process.

Finally, from the experimental data [2] for the dependence of coefficients B and τ on the tapping intensity Γ one can infer that a relation $B = \kappa\tau$ with κ being *independent* of the tapping intensity holds [8]. This implies that only one tap intensity dependent coefficient, the relaxation time τ, should appear in the relaxation law.

Combining the three afore-mentioned ingredients to one single first order differential equation for the rate of change of $A(t)$, one obtains the *minimal relaxation law for the compaction process* under periodic tapping given explicitly by [7]

$$\dot{A} = -\kappa A^2 \exp\left[-\frac{(1-A)}{\kappa\tau A}\right]. \tag{3}$$

Despite the nonlinear character of (3), it is easy to proof that the solution of (3) can be explicitly derived and is determined by

$$A(t) = \frac{1}{1 + \kappa\tau\ln(1 + t/\tau)}, \tag{4}$$

and, therefore, exactly of the functional form (1). The consistency of the asymptotic limits of (3) with the afore-mentioned physical mechanisms for small and large times/tap numbers is also easy to recover. Since the compaction process starts with $A(0) = 1$, the *short time asymptotics* of (3) can be obtained by expanding the exponential term in (3) about $A = 1$. This yields $\dot{A} = -\kappa A^2[1 - (1 - A)/\kappa\tau A +$ h.o.t.] and, as a consequence, (3) reduces to $\dot{A} = -\kappa A^2$ if $1 + \kappa\tau \gg 1/A$ or, equivalently, $t \ll 2\tau$. Moreover, it directly leads to an algebraic decay $A(t) = 1/(1 + \kappa t)$ and, in accordance with the biased void diffusion model, to a relaxation of the packing fraction $\rho(t) \propto 1/t$ for short times, $t \ll 2\tau$. To understand the *long time asymptotics* of (3), it is convenient to recast (3) in the form $\dot{A} = -\kappa \exp(1/\kappa\tau)\exp\left[-(1/\kappa\tau A)(1 - 2\kappa\tau A \ln A)\right]$. The reduced packing fraction $A(t)$ becomes very small in the long time limit and, therefore, (i) $A \ln A$ approaches zero as $t \to \infty$ and (ii) (3) is asymptotically equivalent to $\dot{A} = -\kappa \exp\left[-(1 - A)/\kappa\tau A\right]$ for long times or large tap numbers. This implies that (3) contains the collective reorganization model as long-time limit.

As a consequence, the relaxation law (3) for the reduced packing fraction, *unifies in a natural way* two mesoscopic mechanisms for granular compaction that are dominant on different time scales of the compaction process and are *both* needed to explain the experimental result (1). For short times $t \ll \tau$, biased void diffusion [4] with an algebraic relaxation $\rho(t) - \rho(\infty) \propto 1/t$ dominates granular compaction, whereas the collective reorganization with a relaxation $\rho(t) \propto 1/\ln t$ dominates for long times $t \gg \tau$. The crossover of these effects happens at $t \propto \tau$. In the experiments [2], a strong dependence of the relaxation time τ on the tapping amplitude Γ has been

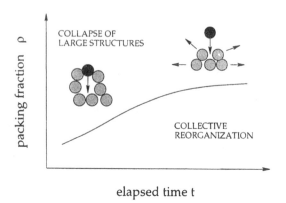

Figure 1: Sketch of the two different mechanisms governing the relaxation behavior of the packing fraction under periodic tapping for short and long times or small and large tap numbers

detected: τ strongly decays from values of about 10^5 for $1.4 < \Gamma < 2$ to values of about 2 for $\Gamma > 3$. This implies that the range of dominant biased void diffusion can vary over several decades. In particular, the biased void diffusion seems to be an essential ingredient for the understanding of compaction for comparatively low tapping amplitudes, $1 < \Gamma < 2$. Moreover, the intuitive expectation that stronger tapping leads to a faster destruction of arches and bridges in a granular system is also well represented in the model (3).

3   Langevin Extension of the Relaxational Law

So far, our considerations solely apply to the dynamics of the packing fraction if averaged over several individual runs. Experimental investigations of one individual compaction process [3] have revealed that the dynamics of the packing fraction fluctuates jerky-like about its ensemble-averaged behavior with characteristic stochastic features in the already strongly compacted state. To mimic the fluctuating dynamics of the packing fraction of an *individual* compaction process, we phenomenologically generalize the relaxational model (3) by the addition of a stochastic or Langevin term $\eta(t)$ yielding

$$\dot{A} = -\kappa A^2 \exp\left[-\frac{(1-A)}{\kappa \tau A}\right] + \eta(t). \tag{5}$$

Here, $\eta(t)$ represent Gaussian white noise fluctuations with zero mean $\langle \eta(t) \rangle = 0$ and a correlation function $\langle \eta(t)\eta(t') \rangle = \Delta \delta(t - t')$, the brackets $\langle ... \rangle$ the ensemble average, and Δ the fluctuation strength.

Since the first term on the right hand side of (5) generally decays rapidly to small values (A close to zero), valuable insights in the long-time dynamics of (5) can be

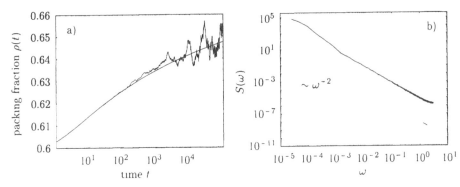

Figure 2: Results of the numerical simulations of (5): a) the relaxation behavior of the packing fraction as a function of time and b) the power spectral density $S(\omega)$ of the density fluctuations. The parameter values are $\kappa = 1/18$, $\tau = 1.4$, and $\Delta = 5 \cdot 10^{-7}$ and correspond to a tapping intensity $\Gamma = 4.5$ in the experiments [2]

obtained from the approximation $\dot{A} = (\rho_0 - \rho_\infty)^{-1}\dot{\rho} = \eta(t)$. This equation determines a Wiener process and can be analysed with elementary methods of stochastics [9]. Its exact solution is determined by $A(t) = A(t_0) + \int_{t_0}^t \eta(t)\,dt$. Assuming that the system is at time $t_0 \gg 1$ in an already well compacted state, one can set $A(t_0) \simeq 0$. This implies a zero average, $\langle A(t) \rangle = 0$, a equal-time correlation function $< A(t)A(t) > \propto \Delta t$, and, therefore, a variance $\langle (\delta A)^2 \rangle = \langle (A(t) - \langle A(t) \rangle)^2 \rangle$ that also increases proportional to Δt in time. Thus, the width of the fluctuations spreads with time as also seen in the experiments [3]. This in turn also justifies the additive coupling of the fluctuations in (5). On the other hand, it also provides a means to estimate the fluctuation strength of the packing fraction in the experimental setup [3]. A detailed comparison with the experimental data [3] shows that Δ should be comparatively small and typically of the order 10^{-6}. In figure 2a) we show the result of a typical numerical integration of the fully nonlinear relaxation model (5) that (i) supports the afore-mentioned argument and (ii) also compares well with the experimental plots presented in [3].

The power spectral density $S(\omega)$ of the fluctuations of the packing fraction is given by $S(\omega) = \langle |\tilde{\rho}(\omega)|^2 \rangle$ with $\tilde{\rho}(\omega) = \int_{-\infty}^{\infty} dt\, \exp(-i\omega t)\rho(t)$ denoting the Fourier transform. Using again the approximation $\dot{A} = (\rho_0 - \rho_\infty)^{-1}\dot{\rho} = \eta(t)$ for an already compacted state and the fact that the power spectral density of $\eta(t)$ is white, one obtains an algebraic decay of the power spectral density of the packing fraction, $S(\omega) \propto \Delta/\omega^2$ with the frequency ω. To substantiate this argument, we have performed numerical simulations of the full nonlinear Langevin model (5) based on the same method of data analysis as described in [3]. After ignoring the first 10^5 time steps or tap numbers, a subsequent time record of 540672 data points has been used to determine the power spectrum $S(\omega)$. A specific example of $S(\omega)$ is depicted in figure 2b) and clearly shows a decay of the power spectrum proportional to $1/\omega^2$. A similar decay behavior has also been detected in the experiments [3] for tapping intensities below $\Gamma = 5$ or, for even larger Γ, at the top of the tube. The model (5), however, cannot reproduce the interesting non-trivial power law behavior at intermediate frequencies ω detected in the experiments [3] for $\Gamma > 5$ and at the bottom of the tube. In that case, the approximation of the fluctuations as Gaussian white noise variables fails.

4   Conclusion

Despite the fact that the compaction of granular matter driven under periodic tapping is micromechanically highly complex, we have reported a comparatively simple, physically based relaxation law that correctly reproduces the experimental results [2]. To study the dynamics of an individual compaction process, we have considered a Langevin generalization of our relaxation model. We have shown that at least two generic features that have been detected in the experiments [3] can be recovered: (i) the broadening of the variance of the packing density fluctuations with time or tap number and (ii) the decay of the spectral statistics for moderate tapping intensities.

References

[1] EWSUK, K.(ed.) (1997): *Compaction Science and Technology.* MRS Bulletin, December issue.

[2] KNIGHT, J.B., FANDRICH, C.G., LAU, C.N., JAEGER, H.M., NAGEL, S.R. (1995): *Density relaxation in vibrated granular material.* Physical Review E. 51:3957-3963.

[3] NOWAK, E.R., KNIGHT, J.B., BEN-NAIM, E., JAEGER, H.M., NAGEL, S.R. (1998): *Density fluctuations in vibrated granular materials.* Physical Review E. 57:1971-1982.

[4] HONG, D.C., YUE, S., RUDRA, R.K., CHOI, M.Y., KIM, Y.W. (1994): *Granular relaxation under tapping and the traffic problem.* Physical Review E, 50:4123-4135.

[5] BEN-NAIM, E., KNIGHT, J.B., NOWAK, E.R., JAEGER, H.M., NAGEL, S.R. (1998): *Slow relaxation in granular compaction.* Physica D, 123:380-385.

[6] BOUTREUX, T., DE GENNES, P.G. (1997): *Compaction of granular mixtures: a free volume model.* Physica A, 244:59-67.

[7] LINZ, S.J., DOHLE, A. (1999): *Minimal relaxation law for compaction of tapped granular matter.* Physical Review E, 60:5737-5741.

[8] LINZ, S.J. (1996): *Phenomenological modeling of the compaction dynamics of shaken granular systems.* Physical Review E, 54:2925-2930.

[9] VAN KAMPEN, N.G. (1992): *Stochastic processes in physics and chemistry*, North-Holland, Amsterdam.

A minimal model for slow dynamics: Compaction of granular media under vibration or shear

Stefan Luding[1] , Maxime Nicolas[2] , Olivier Pouliquen[2]

[1] Institute for Computer Applications 1, Stuttgart, Germany
[2] Groupe Ecoulement de Particules, IUSTI, Technopole de Chateau-Gombert, Marseille, France

Abstract: Based on experiments of the compaction of granular materials under periodic shear of a packing of glass beads, a minimal model for the dynamics of the packing density as a function of time is proposed. First, a random "energy landscape" is created by a random walk (RW) in energy. Second, an ensemble of RWs is performed for various temperatures in different temperature-time sequences. We identify the minimum (mean) of the energy landscape with the maximum (random) density. The temperature scaled by the step-size of the energy landscape determines the dynamics of the system and can be regarded as the agitation or shear amplitude. The model reproduces qualitatively both tapping and shear experiments.

Key-words: Compaction, crystallization, Sinai-diffusion, random walks, quenched disorder

1 Introduction

The issue of slow dynamics, rare events and anomalous diffusion is object to ongoing research in statistical physics [4, 7, 13, 21, 22]. One example for an experimental realization of slow dynamics is the compaction of granular materials, being of interest for both industrial applications and research. Many compaction experiments have been carried out in the last decades, see e.g. [1, 2, 6, 12], but the evolution of density with time is far from being well understood. Recent laboratory experiments concern pipes filled with granular media and periodically accelerated in order to allow for some reorganization [9, 17, 18]. The compaction dynamics was obtained to be logarithmically slow and could be reproduced with a simple parking model [3, 11]. Alternative experiments concern a sheared block of a granular model material (monodisperse glass spheres) and display a similar dynamics [16]. Numerical model approaches, like a frustrated lattice gas (the so-called "Tetris" model) [5, 15, 20] also lead to this slow dynamic behavior, as well as some theory based on stochastic dynamics [14]. The fact that a peculiar dynamics is reproduced by so many models indicates that it is a basic and essential phenomenon.

Rather than modeling granular systems in all details, e.g. in the framework of molecular dynamics or lattice gas simulations [19, 20], we will propose a very simple model based on the picture of a random walk in a random energy landscape, for a review see [4], and even simpler than recent, very detailed considerations in the same spirit [8]. Random walks have been examined, for example, on fractals and ultrametric spaces, where the continuous time random walk was introduced in order to allow a mathematical treatment [10]. If a random walker is situated in a uncorrelated, random, fractal energy landscape, the process is called Sinai-diffusion [4]. The aim of this study is to show that the Sinai model is in qualitative agreement with the compaction dynamics of granular media. An issue not addressed here is a quantitative adjustment which can be reached, for example, by introducing correlations in the EL.

2   Summary of the experimental results

The subject of compaction of granular material [1, 2, 6, 12] has been recently revisited through careful experiments carried out by Knight et al. and Nowak et al. [9, 17, 18]. The experiment consists of a vertical cylinder, full of monodisperse beads, which is submitted to successive distinct taps of controlled acceleration (vertical vibration). The measurement of the mean volume fraction after each tap gives a precise information about the evolution of the compaction. From the first experiments performed with taps of constant amplitude, the increase in volume fraction was found to be a very slow process well fitted by the inverse of a logarithm of the number of taps. Nowak et al. [17, 18] have then studied the compaction under taps of variable amplitude, and showed that irreversible processes occur during the compaction. Starting with a loose packing, the evolution of the volume fraction is not the same when increasing the amplitude of vibration as when decreasing. The first branch appears to be irreversible, whereas the second is reversible.

We have recently performed a compaction experiment based on cyclic shear applied to an initially loose granular packing [16]. A parallelepipedic box full of beads is submitted to a horizontal shear through the periodic motion of two parallel walls at amplitude θ_{max}, see Fig. 1(a). Compaction occurs during this process, leading to crystallisation of the beads in the case of a monodisperse material. The control parameter in this configuration is the maximum amplitude of shear θ_{max} (inclination angle of the walls). The measurement of the mean volume fraction ϕ shows that compaction under cyclic shear is a very slow process as in the vertical vibration experiments (typically 5×10^4 shear cycles or taps). The higher the shear amplitude the more efficient is the compaction (shear amplitudes up to $\theta_{max} < 12.5°$ were examined), when starting with a loose packing of the same initial volume fraction. More surprising results arise when the packing is submitted to a sudden change in shear amplitude. We have observed that a "jump" in volume fraction occurs which is opposite and proportional to the change in θ_{max}. Sudden increase (resp. decrease) in the shear amplitude decreases (resp. increases) the volume fraction (Fig. 1). The

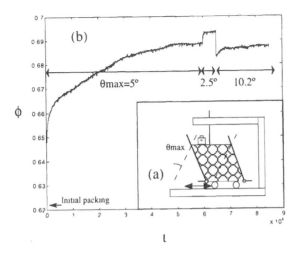

Figure 1. (a) Sketch of the experimental setup. (b) Evolution of the volume fraction as function of shear cycles n: a saturated state is seemingly reached after 6×10^4 cycles with an initial shear angle of 5 degrees: a positive jump in volume fraction is observed when the angle is decreased to 2.5 degrees and applied for 5×10^3 cycles: a negative jump is observed when the angle is increased to 10.2 degrees

response is very rapid (less than 20 cycles) and quasi-independent of the state before the angle change. For more detailed experimental results see [16].

3 The Model

A naive picture that evolves out from the experimental observations is the analogy between the packing of beads and a thermal system seeking for a minimum of energy in a very complex potential-energy landscape [4]. Due to some agitation (shear) compaction occurs and the total potential energy of the packing decreases. Starting from a loose packing at $n = 0$, the vibrational or shear excitation can be seen as the analog of the temperature in the sense that the excitation allows for an exploration of the phase space. In a granular packing of monodisperse spheres, the absolute minimum of the energy is obtained for a perfect (fcc) or (hc) crystal with volume fraction $\phi = 0.74$. If the energy landscape is complex with a lot of different scale valleys or hills, one can understand that an efficient, fast compaction will be obtained with high temperatures: the system is then able to escape the deep local valleys and find a valley with lower potential energy. A decrease of the temperature is then needed to explore local fine-scale minima. The goal of this paper is simply to explore this idea by studying the dynamics of random walkers on a random landscape (the Sinai model), and to show that this very naive picture gives results in qualitative agreement with the experimental observations.

Our model is based on the assumption that all possible configurations of a granulate in a given geometry can be mapped onto an "energy landscape" (EL). Since a simple two-level system (as used to model the dynamics in simple glasses) does not lead to the experimentally observed phenomenology, we assume a fractal energy landscape created by a random walk in energy with phase space coordinate x. A typical EL with energy $V(x)$ is schematically shown in Fig. 2. The stepsize in energy is ΔV, the mean of the energy landscape is V_{mean} and its absolute minimum is V_{min}. Here, the EL is symmetric to its center, in order to allow for periodic boundary conditions in x. Given some EL, the granulate is now modeled as an ensemble of random walkers diffusing on the EL, with a temperature T_{RW}. The analog to the density of a granular packing is the rescaled energy of an ensemble of random walkers in the energy landscape

$$\nu = 1 - \frac{E - V_{\text{min}}}{V_{\text{mean}} - V_{\text{min}}} , \qquad (1)$$

which we denote as *density* in the following. For the (random) initial configuration, the energy will be $E \approx V_{\text{mean}}$ so that $\nu = 0$; for the close-packing configuration, i.e. all RWs are in the absolute minimum, one has $E \approx V_{\text{min}}$ and thus $\nu = 1$. The energy landscape has the size L, the ensemble of RWs consists of R random walkers and S is the number of steps performed.

Since the energy of the RWs in the EL corresponds to the potential energy of the packing, we interpret the (constant) stepsize ΔV (used here as one possibility for the construction of the EL) as a typical activation energy barrier. The maximum of V corresponds to a random loose, local packing density, the mean to the (initial) random close packing, and the absolute minimum to the hexagonal close packing. For one RW, the probability to jump within time Δt from one site x_i to its neighbor-site $x_{i\pm1}$ is

$$p_\pm(x_i) = \min\left[1, \exp(-\Delta_\pm(x_i)/T_{RW})\right] , \qquad (2)$$

with $\Delta_\pm(x_i) = V(x_{i\pm1}) - V(x_i)$. Since ΔV is the only energy scale of the system, we define the dimensionless energy steps $\delta^i_\pm = \Delta_\pm(x_i)/\Delta V$ and the dimensionless temperature $T = T_{RW}/\Delta V$. Written in dimensionless parameters, the jump probabilities are thus $p^i_\pm = p_\pm(x_i) = \min\left[1, \exp(-\delta^i_\pm/T)\right]$, so that a particle always jumps downhill ($\delta^i_\pm \leq 0$), but jumps uphill only with a probability $e_0 = \exp(-1/T)$ (for $\delta^i_\pm > 0$), at finite temperature. The limits $T \to 0$ and $T \to \infty$ correspond thus to immobile particles or to a homogeneous RW, respectively. The discrete master equation for the probability density $n^i(t)$, to find a particle at time t at site i of the EL, is

$$n^i(t + \Delta t) - n^i(t) = -\left[p^i_+ + p^i_-\right] n^i(t) + p^{i+1}_- n^{i+1}(t) + p^{i-1}_+ n^{i-1}(t) \qquad (3)$$

and is (straightforwardly) simulated with $R = 200$ random walkers in an energy landscape with $L = 5000$ sites, if not explicitly mentioned. The time interval Δt

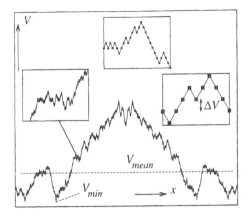

Figure 2: Schematic plot of a typical energy landscape as used for the simulations. The insets show different zoom levels, in order to get an idea of the local situation. For an explanation of the symbols, see the text

corresponds to one Monte-Carlo step but *not* to one shear-cycle n. For the sake of simplicity, we measure x in units of the distance between neighboring sites $\Delta x = x_{i+1} - x_i = 1$, and time in units of Δt. The diffusion constant of a homogeneous random walk ($p_\pm = 1/2$ or $T \to \infty$) is thus $D = \Delta x^2 / \Delta t = 1$. For a constant occupation probability (initial density $n_c = n'(t = 0) = R/L = \text{const.}$), one can extract the diffusion constant D_c as function of the temperature, since $p_\pm = 1/2$ and $p_\pm = e_0/2$ occur with equal probability, so that

$$D_c(T) = \frac{1 + e_0}{2} = \frac{1 + \exp(-1/T)}{2} . \tag{4}$$

In Fig. 3, $D_c = R_2(t)/\sqrt{t}$ is plotted against T after different times t, with $R_2(t) = \langle (x(t) - x_0)^2 \rangle$. For large T, the system does not feel the EL and behaves like an ensemble of homogeneous RWs, whereas its behavior becomes subdiffusive after several steps for small T when the EL is explored. In a situation with $T \approx 0$, after a rather short transient, all RW will occupy a local minimum, a valley ($\vee$), wheras the unstable local configurations hill ($\wedge$), or left- ($/$) and right-slope ($\backslash$) cannot be occupied. if T is small enough, the random walkers stay trapped. Note that this statement is strictly true for $T = 0$ or when a fixed number of shear-cycles or taps is implied. In the Sinai diffusion model, for a finite system, the RW will always find the global minimum – the temperature only determines the time-scale of this process [4]. Since the duration of an experiment is limited, the global minimum cannot be found by all particles if the phase space volume L is large enough.

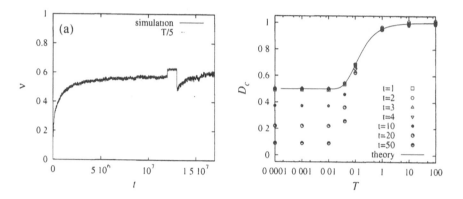

Figure 3: (a) Density plotted against time t (in units of Monte-Carlo steps) for a simulation with $T_0 = 1.1$, $L = 5000$ and $R = 200$. The temperature-sequence is indicated by the dashed line. (b) Diffusion constant $D_c = R_2(t)/\sqrt{t}$ at time t as function of temperature T for systems with constant (initial) density. The solid line shows Eq. (4)

4   Results and Discussion

In parallel to the tapping experiments by Nowak et al. [18], we present simulations of our model using a periodic time series for the temperature. The temperature is kept constant for S steps and then increased by $\Delta T = 0.1$, where it is kept again constant for S steps. T is initially zero and then raised up to $T = 2$. From this state, T is decreased to zero and the loop is repeated seven times. In Fig. 4, the simulation results are displayed for different S as given in the inset. In the initial branch with increasing T, the density increases and slowly decreases for large T. This branch is irreversible, but the periodic loops show almost reversible behavior. For very short loops (small S), the density continuously increases, for longer loops the behavior of the system is reversible. Note that the system also shows hysteretic behavior, the density at decreasing T is below the density at increasing T.

In summary, we presented a very simple model for the dynamics of the compaction of granular media due to an external agitation. Our model is not as detailed as others (parking lot or frustrated lattice-gas models), but it is extremely simple and still shows qualitative agreement with two different types of experiment. Its simplicity allows for future analytical treatments. However, such a simple model arises, besides many others, two major questions: (i) Is it possible to link the real configuration phase space with the an energy landscape? (ii) Is it correct to do the analogy between the experimental external excitations like shear or tapping with a temperature?

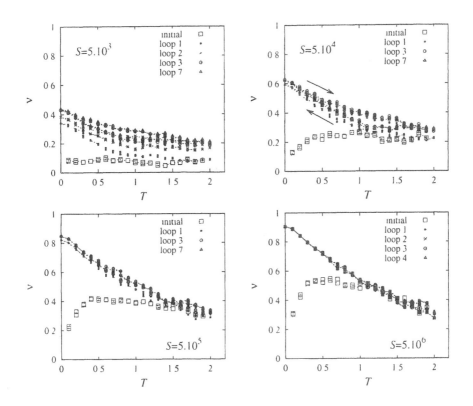

Figure 4: Rescaled density ν plotted as a function of the temperature T for a periodic temperature sequence. The arrows indicate the direction of the loops

Acknowledgements

S. L. acknowledges the support of the I.U.S.T.I., Marseille and thanks for the hospitality. Furthermore, helpful discussions with A. Blumen and I. Sokolov are appreciated.

References

[1] I. Aydin, B. J. Briscoe, and N. Ozkan. Modeling of powder compaction: A review. *MRS Bulletin*, 22(12):1, 1997.

[2] J. E. Ayer and F. E. Soppet. Vibratory compaction II: Compaction of angular shapes. *J. Am. Ceramic Soc.*, 49:207, 1966.

[3] E. Ben-Naim, J. B. Knight, and E. R. Nowak. Slow relaxation in granular compaction. *J. Chem. Phys.*, 100:6778, 1996.

[4] J.-P. Bouchaud and A. Georges. Anomalous diffusion in disordered media: Statistical mechanisms, models and physical applications. *Phys. Repts.*, 195:127–293, 1990. p. 194.

[5] E. Caglioti, V. Loreto, H. J. Herrmann, and M. Nicodemi. A "Tetris-like" model for the compaction of dry granular media. *Phys. Rev. Lett.*, 79(8):1575–1578, 1997.

[6] K. G. Ewsuk. Compaction science and technology. *MRS Bulletin*, 22(12):14–16, 1997.

[7] D. S. Fisher, P. Le Doussal, and C. Monthus. Random walks, reaction-diffusion, and nonequilibrium dynamics of spin chains in one-dimensional random environments. *Phys. Rev. Lett.*, 80:3539, 1998.

[8] D. A. Head and G. J. Rodgers. A coarse grained model for granular compaction and relaxation. *J. Phys. A*, 31(1):107, 1998.

[9] J. B. Knight, C. G. Fandrich, C. N. Lau, H. M. Jaeger, and S. R. Nagel. Density relaxation in a vibrated granular material. *Phys. Rev. E*, 51(5):3957–3962, 1995.

[10] G. Kohler and A. Blumen. Random walks on ultrametric spaces: Mean and variance of the range. *J. Phys. A*, 24:2807, 1991.

[11] A. J. Kolan, E. R. Nowak, and A. V. Tkachenko. Glassy behavior of the parking lot model. *Phys. Rev. E*, 59:3094, 1999.

[12] H. Kuno and J. Okada. The compaction process and deformability of granules. *Powder Technol.*, 33:73–79, 1982.

[13] L. Laloux and P. Le Doussal. Aging and diffusion in low dimensional environments. *Phys. Rev. E*, 57:6296, 1998.

[14] S. J. Linz and A. Dohle. Minimal relaxation law for compaction of tapped granular matter. *Phys. Rev. E*, 60(5):5737, 1999.

[15] M. Nicodemi, A. Coniglio, and H. J. Herrmann. A model for the compaction of granular media. *Physica A*, 225:1–6, 1995.

[16] M. Nicolas and O. Pouliquen. Compaction d'un empilement sous cisaillement périodique. In *14eme congres francais de mecanique*, Toulouse, Septembre, 1999. and in preparation 2000.

[17] E. R. Nowak, J. B. Knight, E. Ben-Naim, H. M. Jaeger, and S. R. Nagel. Density fluctuations in vibrated granular materials. *Phys. Rev. E*, 57(2):1971–1982, 1998.

[18] E. R. Nowak, M. Povinelli, H. M. Jaeger, S. R. Nagel, J. B. Knight, and E. Ben-Naim. Studies of granular compaction. In *Powders & Grains 97*, pages 377–380. Rotterdam, 1997 Balkema.

[19] G. Peng and T. Ohta. Logarithmic density relaxation in compaction of granular materials. *Phys. Rev. E*, 57:829, 1997.

[20] M. Piccioni, M. Nicodemi, and S. Galam. Logarithmic relaxations in a random-field lattice gas subject to gravity. *Phys. Rev. E*, 59:3858, 1999.

[21] S. Rajasekar and K. P. N. Murthy. Dynamical evolution of escape probability in the presence of sinai disorder. *Phys. Rev. E*, 57:1315, 1998.

[22] H. Schiessel, I. Sokolov, and A. Blumen. Dynamics of a polyampholyte hooked around an obstacle. *Phys. Rev. E*, 57:2390, 1997.

7 Numerical investigations:
Distinct element models, lattice models

DEM simulations of uniaxial compression and decompression

Colin Thornton and Lianfeng Liu

School of Engineering and Applied Science, Aston University, Birmingham, UK

Abstract: Numerical simulations of uniaxial compression and decompression of polydisperse systems of elastic and elastoplastic spheres are reported and the results are compared with experimental data obtained from tests on glass ballotini. The results obtained from simulations in which the particles were modelled as elastic spheres exhibited very different unloading behaviour from that provided by the physical experiment; but the stress path followed during loading and unloading was typical of that predicted by cap-type continuum models. By modelling the particles as elastic, perfectly-plastic, qualitative agreement with the experimental data was obtained.

1   Introduction

The basic version of the TRUBAL code [1] used at Aston models the particles as elastic spheres with interparticle friction and adhesion. The interactions between contiguous particles are simulated by algorithms which are derived from theoretical contact mechanics, details of which are provided in [2][3]. More recently the code has been extended to model local plastic deformation at the interparticle contacts[4][5] . Multiaxial compression tests are simulated on a representative volume element, with periodic boundaries, subjected to uniform applied strain fields and the ensemble stress tensor is calculated using statistical mechanics. In this way "perfect" elemental tests are obtained free from boundary effects. Results of three dimensional quasi-static shear simulations of both dense and loose systems of polydisperse systems of elastic spheres have been reported [6]-[10]. Simulations have included both monotonic and cyclic loading tests, constant mean stress tests and constant volume tests, radial deviatoric loading paths, a constant deviatoric strain test and multi-axial plane strain tests. In all cases, the macroscopic behaviour was found to be in good qualitative agreement with results obtained from real experiments on sand.

Having demonstrated that realistically looking macroscopic stress-strain data could be generated by the TRUBAL code, for a wide variety of complex loading paths, it appeared logical that the next step was to examine the quantitative correlation with real experimental data. To this end, it was arranged for experiments to be carried out on glass ballotini in the true triaxial apparatus at Grenoble and the corresponding numerical simulations to be performed at Aston. In all, six experiments were performed in Grenoble. Five of the tests were radial deviatoric loading tests in which the mean stress was maintained constant. The sixth test was a uniaxial compression test which was performed for the specific purpose of calibrating particle and ensemble properties and preliminary simulations of this test were reported in [11]. Here we re-examine the comparison between numerical and experimental results obtained for the uniaxial compression test.

2   Physical experiment

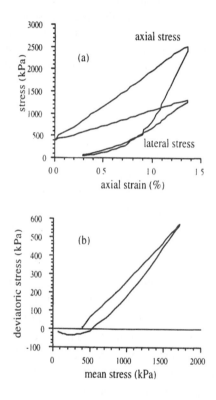

Figure 1: Experimental results (a) stress-strain curves (b) stress path

In the Grenoble true triaxial apparatus a cubical specimen can be subjected to general three-dimensional states of stress by six, kinematically controlled, rigid platens which surround the specimen. The principal stress and strain directions are assumed to remain normal to the platen faces which do not rotate. The three principal stresses are determined from normal pressure sensors mounted in the platens and in contact with the specimen. The three principal strains are determined from linear transducers mounted directly between parallel platens. Details of the equipment design, testing procedures and sample preparation techniques are provided in [12]. Results of previous tests on glass ballotini are reported in [13].

The glass ballotini used in the experiments was nominally $0.1\ mm$ in size and specimens were prepared by spooning and tamping. The uniaxial compression test consisted of three stages (i) isotropic compression to a stress level of $400\ kPa$, (ii) uniaxial compression with "zero lateral strain" until the axial stress was $2500\ kPa$ and (iii) uniaxial unloading with "zero lateral strain" until the axial stress was $50\ kPa$. The results of the uniaxial loading and unloading stages are shown in Figure 1.

Figure 1a shows the evolution of the axial stress (σ_z) and lateral stress ($\sigma_x = \sigma_y$) with axial strain. During the loading stage, both stress-strain curves are essentially linear and the incremental stress ratio, $\Delta\sigma_x/\Delta\sigma_z = 0.43$. During the unloading stage, the axial stress reduces more rapidly than the lateral stress until an isotropic stress state is attained. Thereafter, the variation of both stresses with strain is similar, but with the lateral stress greater than the axial stress since the specimen is now in an axisymmetric extension state. In Figure 1b, the deviatoric stress, $\sqrt{2}(\sigma_z - \sigma_x)/3$, is plotted against the mean stress, $(\sigma_z + \sigma_y + \sigma_x)/3$, to illustrate the stress path followed during loading and unloading.

3   Preliminary numerical simulations

Thornton and Lanier [11] reported computer simulations on a system of 5000 elastic spheres of 5 different sizes within the range $0.1\text{-}0.2\ mm$. This size range was selected on the basis of optical measurements of a sample of the glass ballotini tested in the true triaxial apparatus. The mechanical properties of the particles were specified as: Youngs modulus $E = 70\ GPa$, Poissons ratio $\nu = 0.3$, specific gravity $G = 2.65$ and coefficient of interparticle friction $\mu = 0.3$. The particles were initially randomly generated in a periodic cell and then subjected to an isotropic strain field in order to create a specimen at a stress level of $400\ kPa$. In the six experiments carried out in Grenoble the average packing fraction at a stress level of $400\ kPa$ was ca. 0.7. The initial target was to obtain the same packing fraction in the simulations but this proved to be difficult. A number of different techniques were tried, without prestressing the particle system, but the densest system that could be achieved at a pressure of $400\ kPa$ was only 0.63. Having simulated the isotropic compression stage, two principal strains were set to zero and, in the third principal direction, the system was compressed until the stress in that direction (σ_z) was approximately $2500\ kPa$. Maintaining the zero lateral strain condition, the axial strain was then reversed so that unloading occurred until the axial stress approached zero. The numerically simulated stress-strain data and stress path evolution are shown in Figure 2.

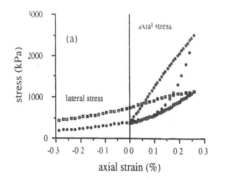
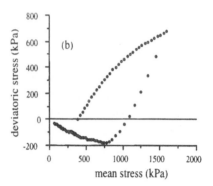

Figure 2: Simulated data, elastic spheres (a) stress-strain curves (b) stress path

It is clear that the simulated system is much stiffer than the physical specimen, even though the initial solid fraction is lower. Due to elastic deformation of the platen assembly, a true zero lateral strain condition cannot be achieved in the laboratory experiment. Measurements indicated that, during uniaxial loading, lateral expansive strains of the order of 0.23% occurred. A second simulation was performed in which the same amount of lateral strain was permitted but this only resulted in an increase in the maximum axial strain from 0.256% to 0.34% which is still much less than 1.36% measured in the laboratory experiment.

A more significant difference is the stress evolution during unloading. As can be seen from Figure 2a, there was only a small rate of decrease in the lateral stress during unloading and, in contrast to the laboratory results, the lateral stress during unloading was higher than the lateral stress during loading. Furthermore, although the initial reduction in axial stress during unloading would appear to be reasonable, it was not possible to reduce the axial stress to a value less than 400 kPa without applying large extensional strains in the axial direction. It is also noted that during unloading the stress state returned to the initial isotropic stress of 400 kPa when the axial strain was zero. However, in spite of these rather unsatisfactory aspects of the simulated results, it is interesting to note that the stress path evolution shown in Figure 2b is remarkably similar to typical predictions by cap-type continuum models.

Thornton and Lanier [11] suggested that the ballotini experienced plastic deformation at the interparticle contacts when under the high stress states applied in the experiments. Consequently, at contacts transmitting large forces in the direction of axial compression [14] the (unloading) contact stiffness during uniaxial decompression would be much stiffer than the (loading) contact stiffness during the previous uniaxial compression stage. It was suggested that this would account for the more rapid reduction in both the axial and lateral stress observed in the experiment.

Based on results of a finite element analysis of a rigid sphere indenting an elastoplastic half-space [15], Thornton [4] suggested that the evolution of the normal contact pressure distribution could be approximated by an "elastic" phase during which the pressure distribution was Hertzian followed by a "plastic" phase during which the pressure distribution was described by a truncated Hertzian pressure distribution by defining a "limiting contact pressure" p_y, Thornton and Ning [5]. Based on these simplifying assumptions it was shown that the force-displacement relationship during "plastic" loading is linear with a normal contact stiffness $k_p = \pi R^* p_y$, with $(\frac{1}{R^*}) = (\frac{1}{R_1}) + (\frac{1}{R_2})$ where R_1 and R_2 are the radii of the two spheres in contact. Unloading was assumed to be elastic but with a reduced contact curvature due to flattening at the contact as a result of the irrecoverable plastic deformation.

Using an alternative version of the computer code which models the interactions between elastoplastic spheres using algorithms based on the above simplified theory, a further simulation of the uniaxial compression/decompression test was reported by Thornton and Lanier [11]. In order to perform the simulation it was necessary to specify a limiting contact pressure p_y which was arbitrarily taken to be $1.5\ GPa$. The results of this simulation of elastoplastic spheres was in reasonable qualitative agreement with the experimental data in so far as both the axial stress and the lateral stress during unloading were less than during loading for a given axial strain and the stress path during the loading-unloading cycle was very similar in shape to the experimentally observed stress path shown in Figure 1b. However, the quantitative agreement was still not satisfactory. Although the axial strain required during loading (0.377%) was an improvement it was still far less than that observed in the experiment.

4   Numerical simulations of systems of elastoplastic spheres

In this section we present results of DEM simulations of a system of 10,000 elastoplastic spheres subjected to the same loading history as the physical experiment reported in Section 2. Nine different particle sizes were used to provide a relatively uniform particle size distribution in the range 100 μm to 316 μm with an average particle diameter of 208 μm. The elastic properties of the particles were $E = 70\ GPa$ and $\nu = 0.3$, as in the previous simulations. In order to examine the effect of the extent of plastic deformation at the contacts, two system were prepared with different values of the limiting contact pressure specified, namely $p_y = 1.0\ kPa$ and $p_y = 0.5\ kPa$. Using an initially low value of interparticle friction ($\mu = 0.01$) each system was isotropically compressed until the mean stress was 400 kPa. At this stage, the solid fraction for both systems was ca. 0.635 which, although significantly lower than that apparently obtained in the Grenoble experiment, is similar to the well known value of 0.637 for random close packing of equal sized spheres. Prior to commencing simulation of the uniaxial compression stage, the interparticle friction was adjusted to the desired value and, for each system, the uniaxial compression/decompression stages of the simulation were repeated using different values of interparticle friction in order to examine its effect on the data obtained.

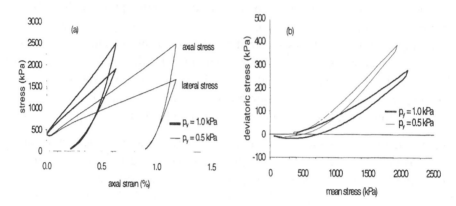

Figure 3: Effect of limiting contact pressure on (a) stress strain curves (b) stress path evolution

Figure 3 shows the results of the simulations of the two systems when the interparticle friction was specified as $\mu = 0.3$. It is clear that the limiting contact pressure has a significant effect on the axial stiffness. The total axial strains of 0.63% (p_y =1.0 kPa) and 1.18% ($p_y = 0.5\ kPa$) are greater than those previously reported by Thornton and Lanier [11] but less than 1.36% observed in the physical experiment. One might consider that the axial stress-strain curve for $p_y = 0.5\ kPa$ is reasonably close to the experimental result. However, the unloading curve is much too stiff when compared with Figure 1a. The incremental stress ratios, $\Delta\sigma_x/\Delta\sigma_z$, of 0.713 ($p_y = 1.0\ kPa$) and 0.60 ($p_y = 0.5\ kPa$) are significantly higher than the apparent experimental value of 0.43. However, these comparisons are further complicated by the fact that the experimental data is not exactly the true material behaviour. Due to elastic deformation of the platen assembly, a true zero lateral strain condition was not achieved in the laboratory experiment. Measurements indicated that, during uniaxial loading, lateral expansive strains of the order of 0.23% occurred [11]. Consequently, the measured lateral stress values are lower than they would be under true zero lateral strain conditions. Furthermore, the reported axial strains do not account for elastic deformation of the platen assembly in the axial direction and, hence, overestimate the true axial strains.

The stress path evolutions during the two simulations are shown in Figure 3b which illustrates the sensitivity to the limiting contact pressure p_y. Both stress paths exhibit reasonable qualitative agreement with the experimental data in Figure 1b.

Figure 4 shows the effect of changing the interparticle friction on the stress- strain curves and the stress path evolution for the system with $p_y = 0.5\ GPa$. Figure 4a indicates that the interparticle friction does not significantly affect the magnitude of the lateral stress developed during uniaxial compression but does significantly affect the axial stiffness. Consequently, the apparent dependence of the incremental stress ratio, $\Delta\sigma_x/\Delta\sigma_z$, on the interparticle friction is entirely due to the fact that the axial stiffness increases with increase in interparticle friction.

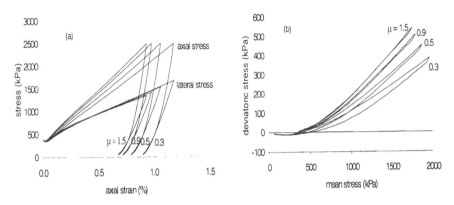

Figure 4: Effect of interparticle friction on (a) stress strain curves (b) stress path evolution

5 Evolution of internal variables

In addition to the macroscopic behaviour discussed above, DEM simulations of granular media provide additional information about the micromechanical behaviour at the particle scale and, consequently, it is possible to examine various internal variables and how they evolve during a simulated experiment. In this section, we consider the induced structural anisotropy, the coordination number and the ratio of contacts sliding at any stage of the simulated experiment.

Structural anisotropy can be characterised by a fabric tensor ϕ_{ij} [16]

$$\phi_{ij} = \frac{1}{C}\sum_1^C n_i n_j = \langle n_i n_j \rangle \tag{1}$$

where n_i defines the components of the unit normal vector at a contact, C is the number of contacts in the representative volume element and $\langle . \rangle$ indicates statistical average. For axisymmetric conditions, the degree of structural anisotropy is defined by the deviator fabric $(\phi_1 - \phi_3)$.

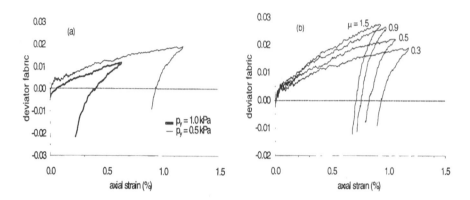

Figure 5: Evolution of structural anisotropy (a) effect of limiting contact pressure (b) effect of inter-particle friction

The evolution of the deviator fabric during the simulations is shown in Figure 5. Figure 5a indicates that the rate at which the induced structural anisotropy develops during uniaxial compresion is not very dependent on the value of limiting contact pressure specified. A significant feature clearly illustrated in the figure is that, from the very beginning of decompression, the deviator fabric reduces until it is zero (isotropic structure) when the stress state is approximately isotropic. With further unloading into axisymmetric extension states of stress there are more contacts developed in the lateral direction than in the axial direction, as indicated by the negative values of the deviator fabric. Qualitatively, the evolution of induced structural anisotropy is not changed by changing the interparticle friction, Figure 5b, but as can be seen in the figure, increasing the interparticle friction increases the degree of structural anisotropy induced during uniaxial compression.

The average coordination number is usually defined as $Z = 2C/N$ where C is the number of contacts and N is the number of particles. However, numerical simulations have revealed that, at any time during shear, there are some particles with no contacts and some particles with only one contact. None of these particles are contributing to the stable state of stress. Hence a mechanical average coordination number is defined as

$$Z_m = \frac{(2C - N_1)}{(N - N_0 - N_1)} \tag{2}$$

where N_1 and N_0 are the number of particles with only one or no contacts respectively. The evolution of the mechanical coordination number is shown in Figure 6. During uniaxial compression the coordination number is greater for the system with a higher value of p_y, Figure 6a. Increasing the interparticle friction reduces

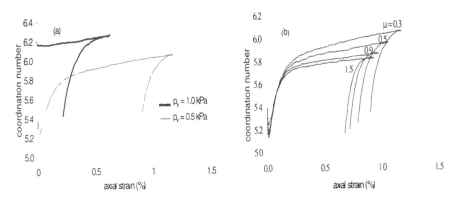

Figure 6: Evolution of coordination number (a) effect of limiting contact pressure (b) effect of interparticle friction

the coordination number, as shown in Figure 6b. In all cases, there is a relatively rapid decrease in the coordination number, i.e. the number of contacts, soon after unloading commences.

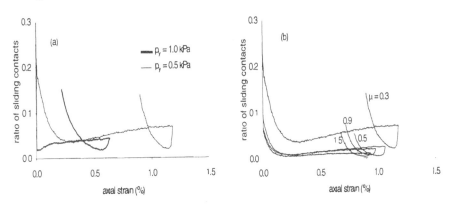

Figure 7: Evolution of ratio of sliding contacts (a) effect of limiting contact pressure (b) effect of interparticle friction

Figure 7 shows how the ratio of contacts which are currently sliding evolves during the simulated experiments. There are not many sliding contacts during uniaxial compression. In all cases, the percentage of contacts sliding is less than 10% and reduces if the interparticle friction is increased. During decompression, there is a brief drop in the ratio of contacts sliding at commencement of unloading but then the ratio of sliding contacts increases continuously with further unloading.

One significant feature demonstrated in Figures 5, 6 and 7 is that, even at the beginning of decompression, the macroscopic behaviour cannot be considered to be elastic according to the physics that occurs at the particle scale. At all stages of unloading, interparticle sliding occurs, interparticle contacts are lost and there is an irrecoverable deformation of the microstructure.

6 Conclusions

Results of DEM simulations of uniaxial compression and decompression of polydisperse systems of elastoplastic spheres have been presented and compared with corresponding experimental data for glass ballotini. Qualitatively, the comparisons between the experimental and simulated data sets are reasonably good. However, the quantitative agreement is not entirely satisfactory.

The numerical simulations demonstrate that the axial stiffness during uniaxial compression is sensitive to both the amount of plastic deformation occurring at the contacts within the assembly of particles (as dictated by the value of p_y chosen) and the interparticle friction. Interestingly, the interparticle friction does not appear to have a significant effect on the magnitude of the lateral stress developed during uniaxial compression.

It has been shown that, at all stages of unloading, interparticle sliding occurs, the number of interparticle contacts reduces and irrecoverable deformation of the microstructure occurs. Consequently, although the initial part of the unloading stress-strain curve may be approximately linear this does not mean that the system response is elastic.

7 Acknowledgement

The authors wish to thank Prof. J. Lanier for providing the experimental data.

References

[1] CUNDALL.P.A.(1988):*Computer simulations of dense sphere assemblies.* In SATAKE, M. & JEMKINS.J.T.(eds.), Micromechanics of Granular Materials:113-123. Amsterdam: Elsevier.

[2] THORNTON, C. & YIN K. K. (1991): *Impact of elastic spheres with and without adhesion.* Powder Technology **65**:153-166.

[3] THORNTON, C. (1999): *Interparticle relationships between forces and displacements.* In M. ODA & K. IWASHITA (eds.), Introduction to Mechanics of Granular Materials, Balkema, 207 - 217.

[4] THORNTON, C. (1997):*Coefficient of restitution for collinear collisions of elastic-perfectly plastic spheres.* J. Appl. Mech. **64**: 383-386.

[5] THORNTON, C. & NING, Z. (1998): *A theoretical model for the stick/bounce behaviour of adhesive, elastic-plastic spheres.* Powder Technology **99**: 154-162.

[6] THORNTON, C. & SUN G. (1993): *Axisymmetric compression of 3D polydisperse systems of spheres.* In THORNTON, C.(ed.), Powders & Grains 93 : 129-134. Rotterdam: Balkema.

[7] THORNTON, C. & SUN G. (1994): *Numerical simulation of general 3D quasi-static shear deformation.* In SMITH I. M. (ed.), Numerical Methods in Geotechnical Engineering, 143-148. Rotterdam: Balkema.

[8] THORNTON, C. & ANTONY, S. J. (1998):*Quasi-static deformation of particulate media.* Phil. Trans. Roy. Soc. London A **356**: 2763-2782.

[9] THORNTON, C. & ANTONY, S. J (2000): *Quasi-static shear deformation of a soft particle system.* Powder Technology (in press).

[10] THORNTON, C. (2000): *Numerical simulations of deviatoric shear deformation of granular media.* Géotechnique **51**: (in press).

[11] THORNTON, C. & LANIER, J. (1997):*Uniaxial compression of granular media: numerical simulations and physical experiment.* In R. P. Behringer & J. T. Jenkins (eds.), Powders & Grains 97, Balkema, 223-226.

[12] LANIER, J. & Z. ZITOUNI (1988): *Development of a data base using the Grenoble true triaxial apparatus.* In SAADA A. & BIANCHINI G.(eds.), Constitutive Equations for Granular Non-Cohesive Soils : 47-58. Rotterdam: Balkema.

[13] LANIER, J. (1976): *Etude expérimentale des lois comportement en grandes déformations á l'aide d'une presse réellement tridimensionnelle.* Cahier du Groupe Francais de Rheologie T.IV, 2: 53-60.

[14] THORNTON, C. (1997): *Force transmission in granular media.* KONA Powder and Particle **15**: 81-90.

[15] HARDY, C., BARONET, C N & TORDION, G. V. (1971): *The elastoplastic indentation of a half-space by a rigid sphere.* Int. J. NUm. Methods Engng **3**: 451-462.

[16] SATAKE, M (1982):*Fabric tensor in granular materials.* In P. A. Vermeer & H. J. Luger (eds.), Deformation and Failure of Granular Materials, Balkema, 63-68.

Generation and effects of different compacted granular materials within the DEM

Markus Herten

Institut für Grundbau, Abfall- und Wasserwesen, Bergische Universität-GH Wuppertal

Abstract: Based on the Distinct Element Method, three dimensional calculations are carried out to determine the effect of different compacted granular materials. Dense, medium dense and loose compactions of sand are examined. The distinct elements with a linear sliding contact model have always the same properties, but different number of contacts in the ball cluster cause varying behaviours. The earth pressure of sand on plane and on circular building pit walls is calculated with respect to their deformation. To define the parameters of the contact model, the well known plane earth pressure according to Coulomb is calculated as a calibration for dense sand. Everything else arises from the porosity of the ball cluster. Simultaneously, model tests were executed to check the results. Despite the simplicity of this DEM which uses only a few parameters, the results of the simulation fit very well to those of the conducted tests and classic theories.

1 Introduction

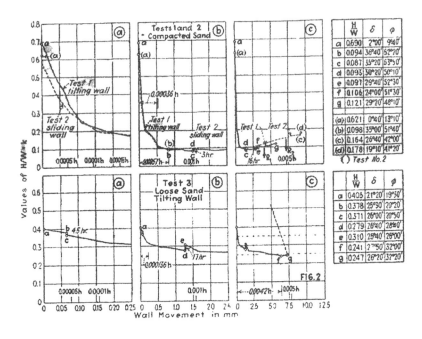

Figure 1: TERZAGHI'S model tests with dry sand [11]

Loose compactions of granular soils have a lower specific weight γ than dense compactions. But additional differences in material properties for example the angle of internal friction φ leads to larger active earth pressure on building pit walls for loose compactions than for dense ones, although the distinct elements i. e. the grains have the same properties.

Already 1934 TERZAGHI [11] published about pressure of dry sand with respect to the movements of the retaining wall. According to the diagrams in figure 1 the active earth pressure coefficient K for the loose sand is about three times bigger than for the dense sand. Nowadays these effects can be simulated with the DEM. Additionally the spatial earth pressure on circular shaft construction is calculated and measured in model tests.

2 Distinct Element Method

The DEM introduced by CUNDALL and STRACK [3] is used to determine the earth pressure arising from radial displacements for different compacted materials. In the DEM the interaction between particles is regarded as a dynamic process that achieves a static equilibrium, when the internal forces are balanced. The dynamic behaviour is represented numerically by a time stepping algorithm. The solution scheme is identical to that used by the explicit Finite Difference Method for continuum analysis. The formal procedure of the DEM takes advantage of the idea to select and define the duration of a time step in such a way that, during a single time step, disturbances in the state of equilibrium can spread only from the regarded particle to its direct neighbours.

The used program PFC^{3D} [9] can be regarded as a simplified application of the DEM in which all bodies are considered to be homogeneous rigid balls. The interaction is described as a soft-contact approach that occurs over an infinite small area. The real deformation of the balls are simulated by overlaps, which leads to limited stiffness. However, the ball stiffness should be selected large enough, so that the deformations of a physical system mainly arise as a result of motions and not out of overlaps.

The calculation presented here uses a linear sliding model for the contacts. It prevents a transfer of normal tensile forces. The shear contact forces are limited, so that any two contact partners can slip on each other. This frictional sliding dissipates the energy contained in the ball system. Usually a local non viscous damping is implemented additionally to dissipate the energy of acceleration. It is only necessary to obtain the initial ball cluster. For the examined deformation process the influence of this damping can be neglected.

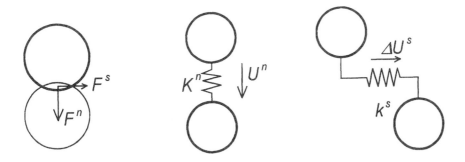

Figure 2: Contact stiffness model [5]

The contacting force vector F is split into a normal force vector F^n and a shearing force vector F^s (Figure 2). The normal contact force is calculated by

$$F^n = K^n U^n n \tag{1}$$

where K^n is the normal stiffness (a function of total displacement and force), U^n is the overlap of the balls and n is the unit normal vector. If the overlap U^n is decreasing to zero, the normal and shear contact forces are set to zero. On the other hand, if the amount of U^n is greater than zero, the contact is checked for sliding. Therefore, the maximum shearing force is determined by

$$F^s_{max} = \mu \left| F^n \right| . \tag{2}$$

If the shearing force is larger, sliding occurs with

$$F^s \leftarrow F^s (F^s_{max} / \left| F^s \right|) . \tag{3}$$

The shear force F^s consists of the increase in the respective time step

$$\Delta F^s = k^s \Delta U^s \tag{4}$$

where k^s is the shear stiffness (relating incremental displacement and force) and of the update of the shear force F^s with respect to the last time step

$$F^s_{new} = F^s_{old} + \Delta F^s . \tag{5}$$

3 Plane Earth Pressure Calculation

To calculate the earth pressure the unknown parameters - normal stiffness k_n, shear stiffness k_s and contact friction μ - must be determined. In this case the normal stiffness is set to $k_n = 10\ 000$ N/m to satisfy the condition concerning the ball overlaps. The other two parameters result from the calculation of the well known plane earth pressure (see figure 1). The methods suggested in [8, 9] (e. g. simulating extension triaxial shear tests) are not applicable, because of the different levels of pressure. Precisely the earth weight neglected in triaxial shear tests is now the main influential factor.

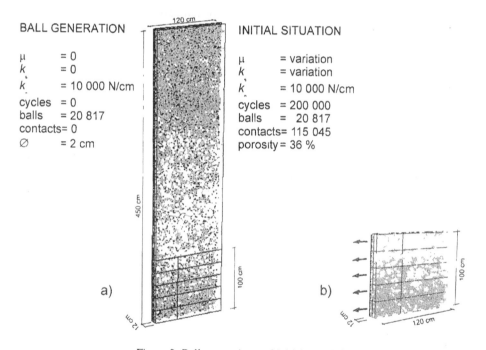

Figure 3: Ball generation and initial state [6]

To obtain a desired ball cluster a number of balls is generated in a space limited by walls of a larger height (Figure 3 a). The diameter of the balls is 2 cm. Initally, there is no contact. By activating the gravitational forces the balls fall downwards and form contacts. The porosity of the ball cluster results from the selected contact friction μ. If the contact friction while spreading μ_e is equal to zero, the minimum porosity of 36 % is achieved. In this case the normal stress on the walls corresponds to the hydrostatic pressure (Figure 3 b).

To obtain the coefficient of earth pressure at rest, K_0, a part of the contact friction μ must be defined. The diameter of the balls is reduced subsequently by a few parts per thousand to activate the internal contact friction. Thus, each stress between hydrostatic pressure $K_y = 1$ and a lower limit value can be adjusted as a function of the contact friction during reduction μ_v.

The procedure to obtain this initial state without any crystal structure takes a long time and computer capacity, but other methods would be useless since they produce an earth pressure coefficient exceeding 1.0 or lead to a regular packing pattern. After a static state is reached the front wall is shifted parallel with a defined constant velocity. The earth pressure decreases from the earth pressure at rest down to a minimum, the active earth pressure that depends on the selected contact friction during sliding μ_s. From this the internal contact friction φ is calculated according to COULOMB. Figure 4 shows the relation between the microscopic contact friction μ_s and the macroscopic contact friction φ for a dense compaction with a porosity rate of 36 %.

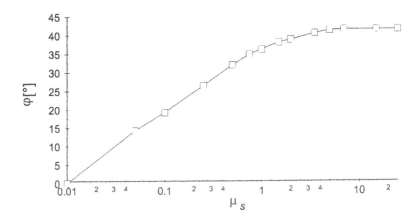

Figure 4: Relation between internal friction and contact friction

To simulate sand with an internal friction of $\varphi = 41.5°$ a contact friction of $\mu_s = 5$ is required. This contact friction appears to be very large, but it should be mentioned that a ball differs in geometry from a sand grain and for that reason a "geometrical" friction has to be added to the physical friction of quartz. With (single) balls and a sliding contact model it is not possible to simulate an internal contact friction exceeding 45° (see figure 4). LORIG et al. [9] uses a contact friction of $\mu = 1.45$ to obtain a internal material friction of about 32°. According to the diagram of figure 1 the active earth pressure of dense sand must be reached at displacements ranging between 0.5 and 1 ‰ of the wall's height. With this condition the shear stiffness k_s is determined. The bottom curve in Figure 5 shows the earth pressure coefficient as a function of the displacement for dense compaction with a contact friction $\mu_s = 5$ and a shear stiffness $k_s = 10\,000$ N/m.

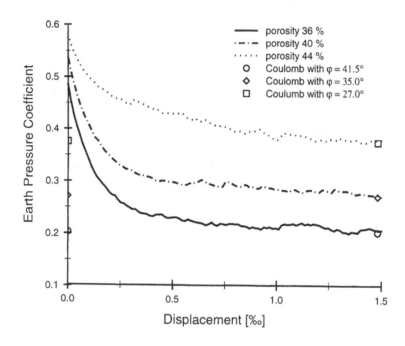

Figure 5: Decrease of plane earth pressure

The middle and the upper curve of Figure 5 present a medium dense and a loose compaction of the ball cluster with porosity rates of 40 % and 44 %. During the shifting process, the parameters are equal to the calculation with dense compaction, but to get more porosity contact friction is defined at the beginning (Table 1). A comparison of the number of balls and contacts turned out that the loose ball cluster with 10 % less balls has of 20 % less contacts. It follows from this that the ratio of contacts per ball also decreases.

Table 1: Parameters for the plane earth pressure calculation

n_0 [%]	μ_e	μ_v	μ_s	Balls	Contacts	$\frac{Contacts}{Ball}$	K_a	φ [°]
36	0	0,364	5,0	20 817	115 045	5,5	0,203	41,5
40	0,36	0,40	5,0	19 319	97 509	5,0	0,271	35,0
44	0,45	1,0	5,0	18 755	91 924	4,9	0,367	27,0

The angle of internal friction for these three degrees of compaction fits very well to the results of the triaxial shear tests for the sand of the model test (Figure 6).

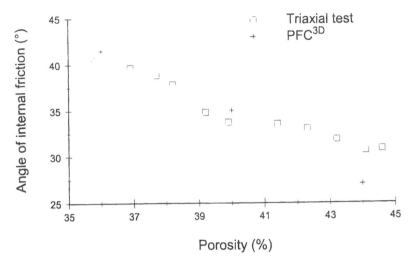

Figure 6: Comparison of triaxial shear tests and calculation

4 Spatial Earth Pressure Calculation

Due to radial symmetry the calculation of the spatial earth pressure can be limited to one sector. Figure 7 depicts a sectional plan and a three dimensional view of the regarded sector. The ratio of shaft height H to diameter D amounts to 2.5. The initial state was produced in the same way as in the plane case.

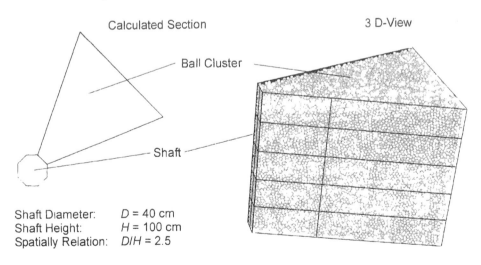

Figure 7: Sectional and 3D-view of the examined ball cluster

A parallel shift of the shaft wall in the radial direction leads to a decreasing earth pressure coefficient for dense compaction as shown by the lower curve in

Figure 8. It is much lower than the plane earth pressure coefficient. The middle and the upper curve show a decreasing earth pressure for medium dense and loose compaction respectively. The achieved final values correspond to the active earth pressure according to WALZ [14] with a tangential arching coefficient K_t between 0.8 and 1.0 and a sliding surface composed of a cylinder and a cone. In contrast to JANNSSEN [7] this classic theory takes the spatial situation into account and is based on TERZAGHI'S [12] reflections on the active earth pressure on shafts.

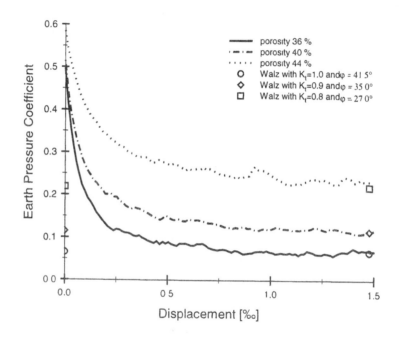

Figure 8: Decrease of radial earth pressure

Table 2 lists the parameters for the three generated ball clusters. The number of contacts decreases again more than the number of balls.

Table 2: Parameters for the spatial earth pressure calculation

n_0 [%]	μ_e	μ_v	μ_s	Balls	Contacts	$\frac{\text{Contacts}}{\text{Ball}}$	K_a	φ [°]
36	0	0,36	5,0	88 570	497 610	5,6	0,065	41,5
40	0,36	0,40	5,0	82 211	421 732	5,1	0,115	35,0
44	0,45	1,0	5,0	79 350	392 122	5,0	0,219	27,0

5 Model Tests

To examine the calculated spatial earth pressure, small scale model tests were carried out. Under the conditions of radial symmetry, it is sufficient to conduct the model tests for only one quadrant of the shaft. Resilient film, lubricant oil and Teflon film offer a solution to decrease the wall friction at the symmetry axis.

The model has a quadratic ground-plan (1 x 1 m) and a height of 1.20 m. A quarter of a cylinder with a diameter of 0.4 m and a height of 1 m is located in one corner and can be moved radially by means of an electric motor. Five measurement plates are fixed at different levels along the shaft height. Dry fine grained sand is filled into the container using the raining technique to achieve a dense or medium dense compactness depending on the height of fall.

6 Comparison

Four tests were executed with dense and medium dense packing (porosity 36 % and 40 %) and the results were compared with the calculated values (Figure 9 and 10). For this purpose, the friction of the shaft wall is taken into account by the simulation

$$\mu_{Wall} = \tan\left(\tfrac{1}{2} \cdot \arctan \mu_{Ball}\right) \tag{6}$$

and by the calculation according to WALZ with $\delta = \frac{1}{2} \cdot \varphi$. The results indicate that even small radial displacements cause a rapid drop in the radial stresses (Figure 11). With dense compaction the calculated and the measured spatial earth pressure on the shaft correspond very well. The results of the model tests with medium dense compaction scatter extremely, but the calculated curve is like the regression of the tests. The drop in the radial stress is not so rapid as with dense compaction.

Although the spatial earth pressure is formally applied to the dimensionless earth pressure coefficient K according to COULOMB´s method, it shows a non-linear distribution over the depth. Figure 11 shows the distribution at the beginning and after 0.5 ‰ shaft displacement for the calculation with PFC^{3D} and for a model test with dense compaction. Additionally the spatial active earth pressure distribution according to WALZ is inserted. Such as the figures beforehand this classic theory leads to an earth pressure that is independent of displacement but corresponds very well with the minimum values at large deflections.

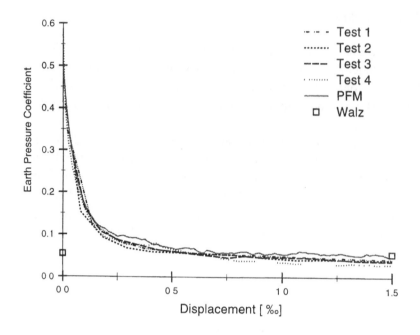

Figure 9: Comparison of the dense compactions

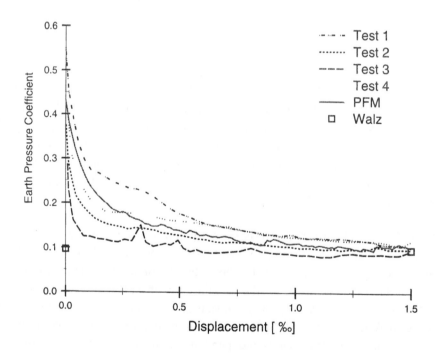

Figure 10: Comparison of the medium dense compaction

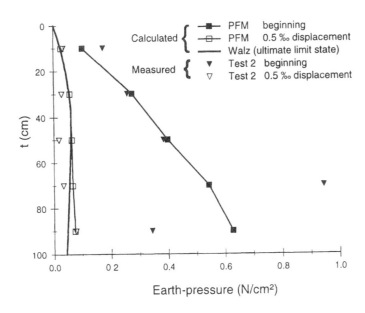

Figure 11: Earth pressure distribution over the depth

7 Conclusion

The good correspondence of calculated and measured results for different compacted soils demonstrates the usefulness of the DEM-procedure: an approximately static situation can be calculated through a dynamic method with a restricted number of parameters. The immense time needed for the calculations - a consequence of the large number of balls and time steps - will become increasingly insignificant over the next years as computer capacity advances further. Therefore, it seems quite obvious that the DEM is a powerful tool to model the behaviour of granular materials in static and dynamic states from a soil mechanic point of view.

References

[1]   BARDET, J.-P. AND PROUBET, J. (1991): *A Numerical Investigation of the Structure of Persistent Shear Bands in Granular media.* Geotechnique 41, No.4, 1: 599-613.

[2]   CUNDALL, P.-A. (1989): *Numerical Experiments on Localization in Frictional Materials.* Ingenieur-Archiv 59, 1:148-159.

[3]   CUNDALL, P.-A. AND STRACK, O.-D.-L. (1979): *A discrete numerical model for Granular assemblies*. Géotechnique 29, 1: 47-65.

[4]   DUBUJET, P. AND DEDECKER, F. (1998): *Micro-mechanical analysis and modelling of granular materials loaded at constant volume*. Granular Matter 1, Springer-Verlag, 1: 129-136.

[5]   HERTEN, M. (1999): *Raumlicher Erddruck auf Schachtbauwerke in Abhangigkeit von der Wandverformung* Bergische Universität GH-Wuppertal, report number 22, Shaker Verlag

[6]   HERTEN, M. AND PULSFORT M. (1999), *Determination of spatial earth pressure on circular shaft construction*. Granular Matter 2, 1-7, Springer Verlag

[7]   JANNSSEN, H.-A. (1895): *Versuche uber Getreidedruck in Silozellen*. Journal VDI 39, 1: 1045.

[8]   LORIG, L.; GIBSON, W.; ALVIAL, J. AND CUEVAS, J.: *Gravity Flow Simulation with the Particle Flow Code (PFC)*. ISMR News J. 3 (1), 1: 18-24.

[9]   PFC^{3D} (1995): *User's Manual*. Itasca Consulting Group, Minneapolis, Minnesota.

[11]   TERZAGHI, K. (1934): *Large Retaining-Wall tests*. Engineering News-Record, February 1

[12]   TERZAGHI, K. (1936): *Distribution of the Lateral Pressure of Sand on the Timbering of Cuts* Proceedings on Intern. Congress of Soil Mechanics and Foundation Engineering.

[13]   VAN BAARS, S.: *Communications on Hydraulic and Geotechnical Engineering*. Discrete Element Analysis of Granular Materials. ISSN 0169-6548 Report No. 95-4 TU Delft Faculty of Civil Engineering.

[14]   WALZ. B.: *Berechnung des raumlich aktiven Erddrucks mit der modifizierten Elementscheibentheorie*. Bergische Universität-GH Wuppertal, report number 6.

Slow relaxation in a diffusional model for granular compaction

Mário J. de Oliveira[1] and Alberto Petri[2]

[1] Universidade de São Paulo, Brazil
[2] Consiglio Nazionale delle Ricerche (CNR), Italy

Abstract:

We investigate the behaviour of a lattice model based on the ideas of random sequential adsorption and diffusional relaxation, for the compaction of a granular system subject to random perturbation and gravity. The lattice is composed by a given number of horizontal layers where non overlapping particles diffuse. Besides diffusion within its own layer a particle suffers a downfall to the layer below whenever there is enough space there.

We restrict at first to the case of one-dimensional layers, and observe, both numerically and by a mean field approximation, an algebraic growth of the particle density that becomes slower and slower as the particle size increases with respect to the lattice spacing, both numerically and by a mean field approximation. We then consider a modification of the model that allows for a nearly exact treatment in the limit of low densities, and show that the same algebraic decay takes place in the general d dimensional case suggesting a logarithmic-like behaviour in the continuum limit.

1   Introduction

The volume fraction of a granular medium poured into a container and subject to shaking or tapping decreases slowly, the origin of this behaviour being the existence of space regions and local configurations that cannot be accessed unless the motion of a large number of grains is involved. Quantitative experimental results suggest a logarithmic increase of this apparent density with time under a periodical perturbation [1], whilst theoretical approaches lead to behaviours ranging from exponential to logarithmic [2, 3].

Density relaxation may depend in principle on many factors, like the geometry of grains, polydispersity, nature of friction forces, etc., and also on the way the system is prepared, i.e. its initial state. As a simplified model we consider a lattice of sites, composed by a given number of horizontal layers in which sites may be occupied by the center of identical particles. By taking the linear size of the particles greater than the lattice spacing excluded volume effects arise and relaxation can be very likely expected to be slower than exponential. We assume here that friction and more complicated steric interactions can be disregarded, so that dynamics is ruled only by inertial and excluded volume effects. Our aim is not to obtain a very realistic

275

model for a granular medium, but to study the compaction process in circumstances that are simple enough for understanding the dynamics of the process.

2   The K-mer model

The K-mer lattice model is intended to describe a system of identical particles whose density increases with time owing to a random motion of the particles along the horizontal direction, and to gravity in the vertical direction [4]. The system is made by a stack of horizontal square lattices, and on each layer non overlapping particles of size larger than the lattice spacing move at random horizontally. Besides diffusion within its own layer, a particle suffers a downfall to the layer below whenever there is enough space there. Periodic boundary conditions are applied in the horizontal direction.

2.1   One dimensional case

The simplest system compatible with the above dynamics is a one-dimensional lattice with identical particles, each one occupying k consecutive sites of the lattice. The particles do not overlap but can diffuse randomly along the horizontal with jumping rate W. The system is initially in the jamming state, i. e. no more particles can be added to the lattice before a rearrangement of the others (because there are at most $k - 1$ consecutive empty sites in the lattice in the initial state). Subsequent diffusion gives rise to the formation of voids of size equal to or larger than k lattice spacings, allowing for the introduction of other particles, that we assume to occur through an ideal reservoir located above the layer. It places a new particle, at a deposition rate R, whenever at least k consecutive empty sites are available. When $R \to \infty$ we get a *Diffusion-Limited Deposition* (DLD) model [5] in which particles are added to the lattice as soon as therequired space is made available through diffusion.

Instead of thinking of diffusing k-mers it is convenient to think in terms of diffusing vacancies, the locations of empty sites. In fact, in order to add a particle to the system it is necessary that k vacancies meet themselves to form a k-*vacancy*. By assuming independence of the probability of site occupation (mean field approximation), if we let ρ_v be the fraction of vacant sites the above requirements yields

$$\frac{d\rho_v}{dt} \propto -\rho_v^k \tag{1}$$

from which it follows

$$\rho_v \sim t^{-\delta_k} \qquad \text{with} \qquad \delta_k = \frac{1}{k-1} . \tag{2}$$

This result shows that this model displays an algebraic relaxation that becomes slower and slower as k increases, turning into a logarithmic-like one when k goes to infinity, i.e. when the continuum limit of the model is taken with vanishing lattice spacing a (ak =constant). It is also remarkable that, in practice, the behavior is very slow for any large k and may become hardly distinguishable from a logarithm.

The algebraic relaxation predicted by the mean field approximation is very well reproduced for large times in numerical simulations. However, exponents given by (2) can be incorrect. In particular, in the case $k = 2$ it is well known that $1/2$ is the right exponent [6], a sign that the process is governed by fluctuations of statistics rather than by averages. A simple explanation comes from the fact that the time t required by two diffusing vacancies to meet is proportional to the square of their relative distance L, and since $L \propto 1/\rho_v$, one gets

$$\rho_v \sim t^{-1/2} . \tag{3}$$

Besides this argument, the case $k = 2$ admits exact solutions in many different circumstances [6].

In order to test the mean field predictions we have performed numerical simulations, obtaining the results reported in the table [4]. It is seen that exponents computed in

Table 1. Exponents of the algebraic increasae of density with time as computed by a mean field approximation (MF) and by numerical simulation

k	2	3	4	5	6	8
δ_k (MF)	1	0.50	0.33	0.25	0.20	0.14
δ_k(numeric)	0.50	0.40	0.35	0.29	0.27	0.25

the two ways are different, with the exception of the case $k = 4$. However there are indications that mean field approximation should hold also for $k > 4$ [5], the discrepancies with numerical values being due to the fact that the asymptotic regime (in which mean field is believed to work) was not really attained in the numerical simulations. We will come again to this point in the next section. In any case, in analogy with mean field results, a slow down of the compaction process takes places as k is increased, suggesting a genuine logarithmic-like behaviour for the continuum.

2.2 Two dimensional case

Let us now turn to the two dimensional case by considering a stack of monodimensional layers. We keep considering that the upper layer obeys DLD, that is one more k-mer is added to it as soon as a k-vacancy is made available by diffusion. The rule is different for the lower layers where in order for a new k-mer to be added not only a void large enough is necessary but, also, a k-mer must occupy the upper layer in exact correspondence with the k-vacancy. This in principle makes dynamics in the lower layers very different, since m-vacancies with $m > k$ can now be formed. No particle can leave the bottom layer.

We have simulated systems of dimers ($k = 2$) up to $10,000$ sites $\times$ 100 layers. We have observed a delay between the initial time and the time at which a sensible increase of density starts to appear, that becomes larger and larger as the number of layers is increased, giving rise to an overall initial slowering in the density change. This fact could be relevant to the logarithm-like behavior observed at short times [1, 3]. However no changes are observed in the large time dependence of the total density. Also with different boundary conditions (e.g. imposing conservation of the number of particles on the top layer) we could not observe any change in the density behavior. We simulated also systems made by trimers,($k = 3$) and observed the same exponent as in one dimension. We therefore expect the exponent no to change also for $k = 4$. The numerical effort required for the case $k = 5$ is much larger than in the one-dimensional case. We simulated lattices made by $2,000$ sites $\times$ 20 layers and by $4,000$ sites $\times$ 100 layers: they seem to indicate that the same exponent as in the one-dimensional case with $R \to \infty$ still holds.

3   The pile model

We consider now a (hyper) cubic d dimensional lattice where vacancies diffuse and annhilate whenever k of them meet at the *same* site. This constitutes a modification of the K-mer model in which vacancies are allowed to diffuse independently on the lattice, and that is to some extent easier to investigate also from the analytical point of view.

The status of the system is completely described once the probability distribution for the number of vacancies at each site is given. By taking periodic boundary conditions in the horizontal direction the translational invariance assures that all the sites are equivalent. The dynamic equations that rule the probability $P_i(t)$ of finding i vacancies on a same site at time t are:

$$\frac{dP_0}{dt} = \sum_{j=1}^{k-1} P_{j,k-1} + \sum_{j=0}^{k-1} P_{j,1} - \sum_{j=1}^{k-1} P_{j,0}$$

$$\frac{dP_i}{dt} = \sum_{j=1}^{k-1} P_{j,i-1} + \sum_{j=0}^{k-1} P_{j,i+1} - \sum_{j=0}^{k-1} P_{j,i} - \sum_{j=1}^{k-1} P_{j,i} \quad (i \neq k-1)$$

$$\frac{dP_{k-1}}{dt} = \sum_{j=1}^{k-1} P_{j,k-2} - \sum_{j=0}^{k-1} P_{j,k-1} - \sum_{j=1}^{k-1} P_{j,k-1} \, ,$$

where $P_{i,j} = P_{i,j}(t)$ are the joint probabilities of finding i and j vacancies at two neighbour site (we assume $W = 1/2$ for simplicity). $P_{i,j}$'s dynamics is ruled in turn by the evolution of the $P_{i,j,l}$'s, the *three sites* joint probabilities, and so on. At the lower level this hierarchy can be truncated in a mean field approximation by assuming that probabilities at neighbouring sites are completely uncorrelated: $P_{i,j} = P_i P_j$. By this way, and by taking into account normalization, $\sum_{i=0}^{k-1} P_i = 1$, the above equations become:

$$\frac{dP_0}{dt} = (1 - P_0)(P_{k-1} - P_0) + P_1$$

$$\frac{dP_i}{dt} = (1 - P_0)(P_{i-1} - P_i) + P_{i+1} - P_i$$

$$\frac{dP_{k-1}}{dt} = (1 - P_0)(P_{k-2} - P_{k-1}) - P_{k-1} .$$

An polynomial solution for this non linear system can be obtained by assuming:

$$P_i(t) = \sum_k c_k^{(i)} t^{-(k\alpha_i)}$$

and requiring the consistency of coefficients. On the other hand it is easy seen that at the lowest order this system simplifies to (1), yielding:

$$P_i \approx t^{-\frac{1}{k-1}} \tag{4}$$

as the leading term in the solution. This implies that within this approximation, density $\rho\ (= P_0)$ increases as $1 - t^{-\frac{1}{k-1}}$ for large times, i.e. it shows the same behaviour as the K-mer model.

It is not known *a priori* wether the mean field approach gives the correct behaviour of density for the whole time range. In any case there are arguments supporting the hypotesis that it works at least in the limit of very large times. This regime corresponds to very low densities of vacancies and one can think that in this case it is possible to determine a typical size L of the lattice in which only k walkers are present. The problem of computing the rate of density increase is thus equivalent to compute the probability rate at which k random walkers meet at the same site on a d dimensional lattice of size L. With periodic boundary conditions the translational invariance of the system makes sufficient to consider the probability that $k - 1$ walkers meet at a given point, e.g. the origin, and the independence of the walkers allows to reduce this problem to that of a single walker moving in a D dimensional space with $D = d(k - 1)$. The real trajectory of each walker (with respect to the reference one) is given by the projection of this motion on different subspaces of dimension d.

The avarage time τ at which the D dimensional walker reaches the origin starting from the point $\mathbf{r_0} = (L/2, L/2, /dots, L/2)$, is known [7]:

$$\tau = \frac{L^2}{2} \quad D < 2$$

$$\tau \;=\; \frac{2L^2}{\pi} \ln L \quad D = 2$$

$$\tau \;=\; \text{const } L^D \quad D > 2 \;.$$

Since this corresponds to the average time rate at which the density of the system decreases of an amount k/L^d, we can conclude that

$$\rho \;\approx\; t^{-d/2} \quad d < \frac{2}{k-1}$$

$$\rho \;\approx\; \frac{\ln t}{t} \quad d = \frac{2}{k-1}$$

$$\rho \;\approx\; t^{-1/(k-1)} \quad d > \frac{2}{k-1}$$

This behaviour strongly supports the the validity of mean field for the K-mer model, above the critical dimension $D_c = d(k-1) = 1$. A logarithmic-like behaviour is expected in both cases when the limit to the continuum is taken.

4   Conclusions

We have observed slow (algebraic) behaviour in two lattice models for the compaction of a granular medium. The decrease of the related exponents with increasing the particles size indicates that a logarithmic increase of density in these models likely takes place in the continuum limit. In the K-mer model, in which non overlapping particles diffuse subject to gravity, the characterizing exponents are numerically found to be the same in one and two dimensions, even if in some cases different from the mean field ones, which however should be hold for very large times and k. This conclusion is strongly supported by the behaviour of the pile model, a modification of the K-mer model. Here the same exponents as in the K-mer model are found, in the limit of low densities, also in the general case of d dimensions and without using any mean field approximation. A more detailed behaviour of this model can be obtained in by solving the evolution equations at a higher level of approximation.

References

[1] Knight J.B., Fandrich C.G., Lau C.N., Jaeger H.M. and Nagel S.R. (1995): *Density relaxation in a vibrated granular material*. Physical Review E. Vol. 51; 3957-3963.

[2] Meta A. and Barker G.C. (1992): *Vibrated powders: Structure, correlations, and dynamics*. Physical Review A, Vol. 45; 3435-3446.

[3] Caglioti E., Loreto V., Herrmann H.J., and Nicodemi M. (1997): *A "Tetris-Like" Model for the Compaction of Dry Granular Media*. Physical Review Letters, Vol. 79: 1575-1578, and refs. therein.

[4] de Oliveira M.J. and Petri A. (1998): *Granular compaction, random sequential adsorption and diffusional relaxation*. Journal of Physics A (Letter to the Editor), Vol. 31: L425-L433.

[5] Nielaba P. and Privman V. (1992): *Random sequential adsorption on a linear lattice; effect of diffusional relaxation*. Modern Physics Letters B, Vol. 6: 533-539.

[6] Eisenberg E. and Baram A (1997): *Diffusional relaxation in random sequential deposition*. Journal of Physics A (Letter to the Editor), Vol. 30: L271-L276, and refs. therein.

[7] Montroll E.W. and Weiss G.H (1965): *Random walk on lattices II*. Journal of Mathematical Physics, Vol. 6, 167-175.

8 Numerical investigations: Continuum models

Influence of pressure level and stress amplitude on the compaction of granular soils

Ivo Herle

Institute of Theoretical and Applied Mechanics, Czech Academy of Sciences, Prague, Czech Republic

Abstract: Numerical optimization of compaction processes is possible if using realistic constitutive models for soils. Hypoplasticity represents such a model which can take into account the influence of pressure and density on the behaviour of granular soils. In order to model realistically also the cyclic deformation, a concept of the so-called intergranular strain is introduced. The densification is investigated for a single homogeneous soil element which can be sufficiently defined with single stress and stretching tensors. A numerical study shows the influence of various deformation modes, pressure level and stress amplitude on the rate of densification and on the work input.

1 Introduction

Soils are compacted by increasing their density. The densification results from stress and deformation changes which are imposed on the finite soil volume. This finite soil volume can be considered as composed from soil elements characterized by homogeneous stress and deformation fields.

It can be observed that the densification methods noticeably differ in their efficiency. Starting from the same initial density, certain stress/strain paths reach a final higher density in shorter time and consuming less energy than others. The aim of this paper is to investigate numerically the densification of a single homogeneous soil element. A sandy soil will be assumed the mechanical behaviour of which can be sufficiently described with a hypoplastic constitutive model with intergranular strains. Various deformation modes will be compared and in case of cyclic simple shearing the influence of the pressure level and the stress amplitude will be studied. From the results of the numerical calculations some conclusions concerning the optimum way of densification will be drawn.

2 Constitutive model

2.1 Hypoplastic model

Hypoplastic constitutive models [4] are well suited for the mathematical description of the mechanical behaviour of granular materials as e.g. sands. A hypoplastic model is embedded in a single tensorial evolution equation of the efective stress tensor $\mathbf{T}$ and it takes into account the influence of the mean pressure p_s and of the void ratio e in a wide range of their values [1]. The calculated behaviour depends also on the deformation direction given by the stretching tensor $\mathbf{D}$ (but not on the value of $\|\mathbf{D}\|$ as the equation is rate-independent) and on the induced stress anisotropy stored in $\mathbf{T}$. One does not distinguish between elastic and plastic deformation, and different stiffnesses for 'loading' and 'unloading' are achieved without any explicit switch function. In sequel, the formulation of the hypoplastic equation after [6] is used.

The evolution of e is bounded by the pressure-dependent maximum and minimum void ratios e_i and e_d and in case of monotonic shearing it tends to the critical value e_c:

$$\frac{e_i}{e_{i0}} = \frac{e_c}{e_{c0}} = \frac{e_d}{e_{d0}} = \exp\left[-\left(\frac{3p_s}{h_s}\right)^n\right] \quad . \tag{1}$$

e_{i0}, e_{c0} and e_{d0} are values of the limit void ratios at zero pressure, h_s denotes the so-called granulate hardness and the exponent n is an additional constant.

Using e_d and e_c, the pressure-dependent relative density

$$D_p = \frac{e_c - e}{e_c - e_d} \tag{2}$$

can be defined [1] contrary to the common definition of relative density

$$D_r = \frac{e_{max} - e}{e_{max} - e_{min}} \tag{3}$$

with e_{max} and e_{min} being pressure-independent maximum and minimum void ratios according to various standards.

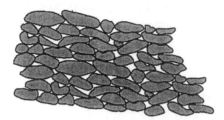

Figure 1: Grain skeleton as a dry masonry

The desirable densification should reach the limit value e_d. It is assumed that for $e < e_d$ the behaviour of a grain skeleton approaches the one of a dry masonry (figure 1). It is however obvious that under practical circumstances the limit of e_d cannot be usually reached. The value of e_d can however be surpassed numerically which can result in computational problems (see Appendix).

The practical experience shows that a monotonic compression does not produce so good densification like a cyclic deformation. This observation is also incorporated in the hypoplastic model, see figure 2. Monotonic compression starting from $e = e_{min}$ departs from e_d curve. This results in decreasing D_p with increase of the mean pressure. On the contrary, cyclic deformation starting from $e > e_{min}$ is able to reach e_d after a finite number of cycles. Cyclic densification can also be achieved with shearing without changing p_s.

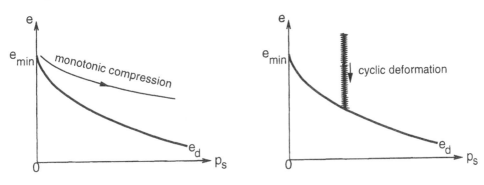

Figure 2: Monotonic (left) versus cyclic (right) densification

2.2   Intergranular strains

Numerical simulation of a monotonic densification can be realistically modelled with the hypoplastic model introduced in the preceding section. In this case, the state of the material is sufficiently characterized by $\mathbf{T}$ and e for a particular deformation direction. Considering a cyclic deformation, additional state variable must be introduced otherwise the so-called ratchetting effect arises, see figure 3. The stress states in point a and b, respectively, are identical and the void ratios differ only insignificantly. In spite of that the observed response during reloading (point b) is much stiffer than during the primary loading. This behaviour cannot be modelled using only $\mathbf{T}$ and e as state variables.

The stiff response after the strain rate reversal results mainly from deformation of grain contacts. To take this effect into account, the so-called *intergranular strain* tensor was proposed [5]. This tensor stores the recent deformation direction and its evolution depends solely on its actual value and the stretching tensor $\mathbf{D}$. The calculated hypoplastic stiffness increases if the 'directions' of the intergranular strain tensor and of $\mathbf{D}$ are different.

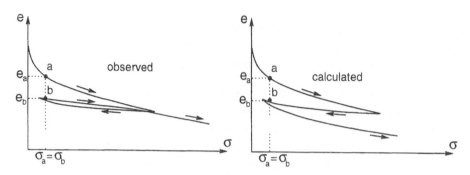

Figure 3: Observed and calculated (without intergranular strains) behaviour in cyclic isotropic compression (schematically). σ denotes the compressive stress

3   Compaction simulation

Numerical simulations were performed with the parameters for Zbraslav sand which are summarized in table 1. The sand characterization and the determination of the hypoplastic parameters are described in detail elsewhere [3]. It has been shown that the hypoplastic parameters are closely related to the index properties of grain assemblies and hence they can be estimated from few simple experiments.

Table 1: Hypoplastic parameters of Zbraslav sand

φ [°]	h_s [MPa]	n	e_{d0}	e_{c0}	e_{i0}	α	β
31	5700	0.25	0.52	0.82	0.95	0.13	1.00

The parameters of intergranular strain are given in table 2. They were taken from [5]. No attempt was made to calibrate the values of table 2 for Zbraslav sand. Consequently, all presented calculations can be considered as qualitative only.

Table 2: Intergranular strain parameters

R	m_R	m_T	β_r	χ
$1 \cdot 10^{-4}$	5.0	2.0	0.50	6.0

Only a single homogeneous soil element with uniquely defined **T**, e and **D** was considered. Therefore a numerical integration of the constitutive equation was sufficient. Details of the procedure for element test calculations are described in [2].

The initial state was assumed close to the in situ conditions with T_{11} denoting the vertical stress and $T_{22} = T_{33} = K_0 T_{11}$ the horizontal stresses with $K_0 = 0.5$. Investigating the role of the deformation mode, the same initial vertical stress $T_{11} = -100$ kPa and the initial void ratio $e = e_c = 0.759$ were used (compressive stresses are negative). Four different stretching tensors **D** were applied which can be described as

1. *iso* – isotropic deformation: $D_{11} = D_{22} = D_{33} \neq 0$, otherwise zero

2. *oed* – one-dimensional compression: only $D_{11} \neq 0$

3. *tx* – triaxial compression: $D_{11} \neq 0$, $D_{22} = D_{33} \neq 0$ and $T_{22} = T_{33} = \text{const.}$

4. *ssh* – simple shear: $D_{11} \neq 0$, $D_{21} \neq 0$ and $T_{11} = \text{const.}$

The cycles were stress-controlled keeping the stress amplitude constant during the calculation. Each cycle consisted of a stress loop from σ to $\sigma + \Delta\sigma$ and back to σ. The calculation finished when $e = e_d$.

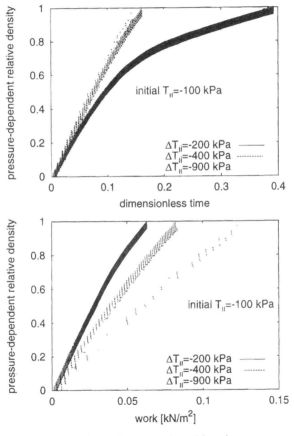

Figure 4: Comparison of isotropic compaction with various stress amplitudes

Figure 4 depicts the results for isotropic compaction using different stress amplitudes. It can be seen that increasing the stress amplitude the compaction time decreases (the time can be considered as a dimensionless parameter only because the equation is rate-independent). However, the work input needed for reaching e_d increases by increasing the stress amplitude (the work input is defined as $\int_t \text{tr}[\mathbf{T} \cdot \mathbf{D}]$).

Although initially K_0 stress state was prescribed, the stress becomes isotropic after several stress cycles. Similar asymptotic stress behaviour was also observed for other deformation modes. This property documents the robust asymptotic behaviour of the hypoplastic model.

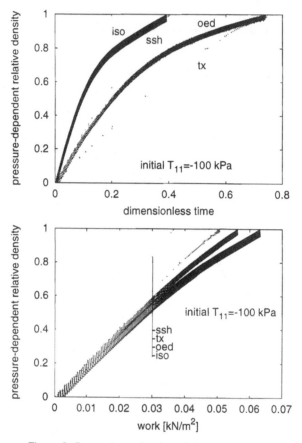

Figure 5: Comparison of various deformation modes

Comparison of various deformation modes is given in figure 5. Different stretching directions require different stress amplitudes which makes the direct comparison somewhat difficult. Nevertheless, from the upper figure it can be seen that the same compaction time was needed for iso and ssh tests and for oed and tx tests, respectively. The lower figure shows that the lowest work was consumed in case of simple shearing (ssh) contrary to the highest work consumption in isotropic compaction. This numerical result coincides well with experimental observations: it is much easier to compact granular soils with cyclic shearing than with cyclic compression.

Further numerical studies are restricted to the compaction due to cyclic simple shearing. The influence of the initial stress can be seen in figure 6. The stress amplitude was kept constant. As can be expected, higher stresses result in higher work input (figure 6 right). Concerning the time (figure 6 left), however, the shearing at higher stresses may take considerably longer time if the stress amplitude is much smaller than the mean pressure (cf. initial $T_{11} = -400$ kPa and $\Delta T_{12} = \pm 50$ kPa). This follows from the concept of intergranular strains: for small strain (i.e. also stress) amplitudes the deformation is limited mainly to grain contacts, there are only minor grain rearrangements and the densification can take a very long time.

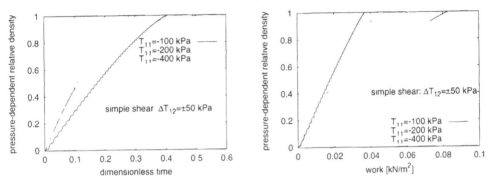

Figure 6: Simple shearing with constant amplitude of the shear stress

In practice, compaction is often controlled by displacements (e.g. vibrator eccentricity) rather by stresses. The results of numerical simple shearing with constant strain amplitude starting at different initial stresses are summerized in figure 7. Whereas the initial stress does not influence the compaction time, the work input noticeably increases by increasing the initial stress.

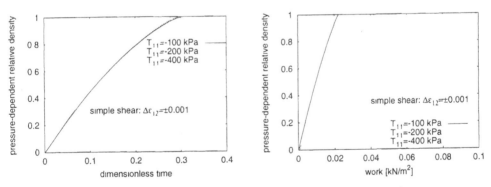

Figure 7: Simple shearing with constant amplitude of the shear strain

Very important phenomena can be revealed in case of different strain amplitudes keeping the same constant vertical stress (figure 8). Choosing the strain amplitude too large, the maximum densification cannot be achieved! After a certain densification the granular soil becomes stiffer and also the limit stress ratio increases. Consequently, the initially purely contractant behaviour changes into the dilatant one and the soil cannot be compacted any more if keeping the strain amplitude constant (figure 9).

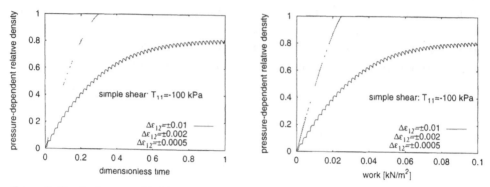

Figure 8: Simple shearing with different strain amplitudes at constant vertical stress $T_{11} = -100$ kPa

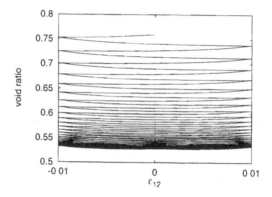

Figure 9: Simple shearing without reaching e_d due to the too large strain amplitude

4 Conclusions

Having a realistic constitutive model, it is possible to look for the optimum compaction modes numerically. This is advantageous for fast testing of many different deformation directions, initial pressures and stress/strain amplitudes. Such an investigation may help in design of compaction methods and devices.

Using the hypoplastic constitutive model it is possible to perform such numerical experiments. Moreover, a close link between the model constants and the index properties of grain assemblies enables a rapid calibration of the model for a particular soil. However, the characterization of the soil state with the effective stress tensor $\mathbf{T}$ and the void ratio e is not sufficient in case of cyclic deformation. Additional state variable like intergranular strain tensor is needed for tracing the recent deformation history.

The results of the calculations presented in this paper suggest that

- the best compaction can be achieved in simple shearing with constant load whereas the worst one follows from cycles of isotropic stress,

- if the ratio of the stress amplitude and the initial stress is too low, the compaction can take rather a long time,

- if the strain amplitude in simple shearing is too large, maximum compaction cannot be reached due to the soil dilation.

Although these findings need not to be recognized as new ones from the practical point of view, it is important to realize that the succesful numerical modelling opens new horizonts for the compaction design and control.

5 Appendix

Using the hypoplastic constitutive model [1], the actual void ratio e must not become lower than the minimum void ratio e_d (1). The performance of the hypoplastic equation in case of $e < e_d$ may be unpredictable. Unfortunately, this constraint must be controlled numerically because the hypoplastic equation cannot guarantee $e \geq e_d$ during the deformation process, particularly for cyclic deformation.

Let be considered the hypoplastic equation without intergranular strains and neglected the corotational terms. In case of $e = e_d$ the factor f_d vanishes and the equation reduces to

$$\dot{\mathbf{T}} = f_b f_e \frac{1}{\text{tr}(\hat{\mathbf{T}} \cdot \hat{\mathbf{T}})} \left[F\mathbf{D} + a^2 \hat{\mathbf{T}} \text{tr}(\hat{\mathbf{T}} \cdot \mathbf{D}) \right] \tag{4}$$

(the functions f_b and f_e depend on p_s and e and their representation can be found e.g. in [6], the constant a is related to the critical friction angle φ_c and F is a function of **T**). Looking at the evolution of the mean pressure $\dot{p}_s = -\text{tr}\dot{\mathbf{T}}/3$ (note that $\text{tr}\hat{\mathbf{T}} = 1$) and assuming the initially isotropic stress state with $F = 1$ and $\text{tr}(\hat{\mathbf{T}} \cdot \mathbf{D}) = \text{tr}\mathbf{D}/3$, (4) can be written as

$$\dot{p}_s = -\frac{1}{3} f_b f_e \frac{1}{\text{tr}(\hat{\mathbf{T}} \cdot \hat{\mathbf{T}})} \left(1 + a^2/3\right) \text{tr}\mathbf{D} = -C \, \text{tr}\mathbf{D} . \tag{5}$$

C is always positive for allowed range of values of the hypoplastic constants. Therefore, increasing the mean pressure results in decreasing the void ratio as expected.

Rewriting (1) as an evolution equation of p_s,

$$\dot{p}_s = -\frac{1}{3}\frac{h_s}{n}\left(\frac{3p_s}{h_s}\right)^{1-n}\left(\frac{1+e_d}{e_d}\right)\mathrm{tr}\mathbf{D} = -C^*\,\mathrm{tr}\mathbf{D}\,, \tag{6}$$

it is possible to compare

$$\begin{aligned}
C &= f_b f_e (1 + a^2/3) = \\
&= \frac{h_s}{n}\left(\frac{3p_s}{h_s}\right)^{1-n}\left(\frac{1+e_i}{e_i}\right)\left(\frac{e_i}{e_d}\right)^{\beta}\left[3+a^2-a\sqrt{3}f_{d0}\right]^{-1}(1+a^2/3)\,,
\end{aligned} \tag{7}$$

using the abbreviation $f_{d0} = \left(\dfrac{e_{i0}-e_{d0}}{e_{c0}-e_{d0}}\right)^{\alpha}$, with

$$C^* = \frac{1}{3}\frac{h_s}{n}\left(\frac{3p_s}{h_s}\right)^{1-n}\left(\frac{1+e_d}{e_d}\right)\,. \tag{8}$$

The exponents α and β are additional soil constants.

The condition $e \geq e_d$ is violated if $1/C^* < 1/C$, i.e. if for the same stress rate the compressive volumetric strain rate is larger after (5) than after (6),

$$\frac{1}{3}\frac{1+e_d}{e_d} > \frac{1+e_i}{e_i}\left(\frac{e_i}{e_d}\right)^{\beta}\left[3+a^2-a\sqrt{3}f_{d0}\right]^{-1}(1+a^2/3)\,. \tag{9}$$

This unequality can be simplified assuming $\beta = 1$ and $f_{d0} = 1$ (both values are within the range of the parameters validity):

$$1 + e_d > (1+e_i)\left(\frac{3+a^2}{3+a^2-a\sqrt{3}}\right)\,. \tag{10}$$

It can be seen that the result depends on a, e_i and e_d. A common critical friction angle for sand $\varphi_c = 30°$ corresponds to $a = 3$. Subsequently, $(1 + e_d)/(1 + e_i) < 1$ if e should remain above e_d. The latter relation is always fulfilled for any realistic values of e_d and e_i. Consequently, the condition $e \geq e_d$ cannot be violated by increasing the mean pressure from the isotropic stress state for any deformation direction. However, (5) and (6) are linear in $\mathbf{D}$ and thus $e \geq e_d$ is immediately violated in case of decreasing p_s, see also figure 10.

Acknowledgement

The financial support by the Grant Agency of the Czech Republic (Grant No. 103/99/P005) is gratefully acknowledged.

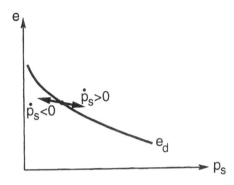

Figure 10: Violation of the condition $e \geq e_d$ for decreasing p_s at $e = e_d$

References

[1] GUDEHUS, G. (1996): *A comprehensive constitutive equation for granular materials.* Soils and Foundations, 36(1): 1–12.

[2] HERLE, I. (1997): *Hypoplastizität und Granulometrie einfacher Korngerüste.* Veröffentlichung des Institutes für Bodenmechanik und Felsmechanik der Universität Fridericiana in Karlsruhe, Heft 142.

[3] HERLE, I., GUDEHUS, G. (1999): *Determination of parameters of a hypoplastic constitutive model from properties of grain assemblies.* Mechanics of Cohesive-Frictional Materials, 4(5):461–486.

[4] KOLYMBAS, D. (1991): *An outline of hypoplasticity.* Archive of Applied Mechanics, 61: 143–151.

[5] NIEMUNIS, A., HERLE, I. (1997): *Hypoplastic model for cohesionless soils with elastic strain range.* Mechanics of Cohesive-Frictional Materials, 2(4):279–299.

[6] VON WOLFFERSDORFF, P.-A. (1996): *A hypoplastic relation for granular materials with a predefined limit state surface.* Mechanics of Cohesive-Frictional Materials, 1(4):251–271.

Settlement of liquefied sand after a strong earthquake

V. A. Osinov and I. Loukachev

Institute of Soil Mechanics and Rock Mechanics,
University of Karlsruhe, D-76128 Karlsruhe, Germany

Abstract: The paper presents the results of the numerical modelling of the liquefaction and subsequent settlement of a saturated sand layer induced by a large-amplitude shear wave coming from below. The repeated shear deformation caused by the dynamic disturbance results in the reduction of the effective pressure and liquefaction of the soil. The permanent changes in the effective stress are inhomogeneous. The liquefaction is localized in thin zones which emerge successively from the top to the bottom and divide the layer into several parts. Each zone of liquefaction isolates the above situated soil from the incoming wave. The settlement of the sand is found from the solution of the quasi-static problem of uniaxial compression. The settlement is presented as a function of the amplitude of the dynamic disturbance. Both the dynamic and the quasi-static problems are solved with a constitutive equation of hypoplasticity.

1   Introduction

A dynamic impact on a granular soil caused by a strong earthquake or an explosion usually leads to the densification of the soil. The dynamic densification proceeds in different ways depending on whether the soil is dry or saturated. A dry soil is compacted immediately during the dynamic disturbance, while the immediate densification of a saturated soil is inhibited by the presence of water. The densification and settlement of a saturated soil proceed slowly as the water seeps out and the water pressure regains its original distribution. Densification of a saturated soil can therefore take place only if the water pressure is increased during the dynamic process so that the state of the soil is changed towards liquefaction. Such changes are observed when a saturated granular soil is subjected to cyclic loading [1,2].

The present paper deals with the numerical modelling of liquefaction and subsequent densification of saturated sand induced by plane shear waves. Correspondingly, the problem is solved in two stages. The first stage consists in the solution of the dynamic problem of shear deformation of a saturated sand layer. The cyclic deformation of the sand during the dynamic disturbance leads to a build-up of the pore water pressure and to the corresponding reduction of the effective stress. The distribution of the permanent stresses resulting from the solution of the dynamic problem is then used at the second stage to solve the quasi-static problem of consolidation. Both the

297

dynamic and the quasi-static problems are solved with a constitutive equation of the hypoplasticity theory.

2   Constitutive equation

By Terzaghi's effective stress principle, the total stress in a fully saturated soil is the sum of the effective stress $\mathbf{T}$ and the isotropic stress $-p_l\mathbf{I}$, where $p_l > 0$ is the water pressure, $\mathbf{I}$ is the unit tensor (compressive stresses are negative). The mechanical behaviour of the granular skeleton is considered to be independent of the pore water pressure.

According to the conception of hypoplasticity [3–5], the constitutive behaviour of a (dry) granular skeleton is described by an equation that establishes a relationship between the stress rate $\dot{\mathbf{T}}$ and the rate of deformation $\mathbf{D}$:

$$\dot{\mathbf{T}} = \mathbf{H}\left(\mathbf{T}, \mathbf{D}, e, \boldsymbol{\delta}\right), \tag{1}$$

where $\mathbf{D}$ is the stretching tensor with rectangular Cartesian components

$$D_{ij} = \frac{1}{2}\left(\frac{\partial v_i}{\partial x_j} + \frac{\partial v_j}{\partial x_i}\right). \tag{2}$$

The tensor-valued function $\mathbf{H}$ in (1) involves the stress tensor $\mathbf{T}$ and the void ratio e as well. In an extended version of hypoplasticity [6], the function $\mathbf{H}$ also includes the so-called intergranular-strain tensor δ which is calculated by the integration of the evolution equation

$$\dot{\boldsymbol{\delta}} = \mathbf{F}\left(\mathbf{D}, \boldsymbol{\delta}\right). \tag{3}$$

The tensor δ carries the information about the history of deformation of the material element and determines the response of the material together with the current stresses and the void ratio. The functions $\mathbf{H}$ and $\mathbf{F}$ in (1) and (3) are nonlinear. They are homogeneous of degree one in $\mathbf{D}$ so that the behaviour of the material is rate independent. Owing to the incremental nonlinearity, the hypoplasticity equation can well describe the change of stiffness under changes of direction of deformation; this equation also describes hysteresis, pressure dependence of stiffness, dilatancy and contractancy for a wide range of pressures, densities and deformations. The constitutive parameters of a particular soil are state independent and can be evaluated on the remoulded samples in the laboratory [7,8]. The detailed representation of the functions (1), (3) and the constitutive parameters of Hochstetten sand which were used in the present study can be found in [5,6].

Propagation of large-amplitude transverse waves in a saturated soil is accompanied by repeated shear of the soil. A granular skeleton tends to contract when subjected to repeated shear deformation. Under undrained conditions, contraction of the skeleton is prevented by low compressibility of the pore water. As a consequence, the effective pressure decreases with each cycle of the deformation. The stiffness of the skeleton decreases as well and tends to zero as the effective pressure vanishes, which results in the liquefaction of the soil.

Liquefaction under repeated deformation is adequately modelled by the extended version of the hypoplastic equation developed in [6]. Figure 1 shows an example of cyclic shear with constant volume calculated by the integration of equations (1) and (3), starting from a hydrostatic stress state $T_{11} = T_{22} = T_{33} = -50$ kPa with a void ratio of 0.8. The figure shows the stress components T_{12} and T_{11} (T_{22} and T_{33} are equal to T_{11} in this case) versus the shear deformation $\gamma_{12} = \partial u_2 / \partial x_1$, where u_2 is the displacement component. The cyclic shear results in the reduction of the effective stress to zero and thus in the full liquefaction of the sample. The rate of liquefaction depends strongly on the amplitude of shear: the higher the amplitude, the less cycles are needed to reach liquefaction. The number of cycles grows rapidly with decreasing amplitude.

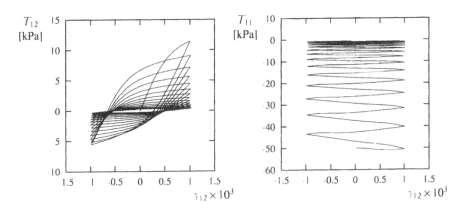

Figure 1: Repeated pure shear calculated for Hochstetten sand.

3 Dynamic problem

The duration of the motion of the soil in the dynamic problem is in the range of few seconds. For a layer of fine sand several metres thick, such times may be considered short enough for the seepage of the pore water to be neglected. The motion of a saturated sand is then described by a single velocity field **v** which is common for both liquid and solid phases. With incompressible grains and water, the velocity satisfies the incompressibility condition

$$\text{div }\mathbf{v} = 0. \tag{4}$$

The equation of motion is written as

$$\text{div }\mathbf{T} - \text{grad }p_l + \varrho\,\mathbf{g} = \varrho\,\dot{\mathbf{v}}, \tag{5}$$

where $\mathbf{g}$ is the mass force vector, $\dot{\mathbf{v}}$ is the material time derivative of the velocity vector, $\varrho = (\varrho_l e + \varrho_s)/(1 + e)$ is the mean density, ϱ_l and ϱ_s are the densities of the liquid and solid fractions respectively.

Consider a layer of fully saturated sand lying on a hard base (rock) as shown in Figure 2. The layer is initially at rest. The initial water pressure and the effective stresses vary with the depth according to the density and the gravitational and buoyancy forces. In what follows we will consider plane-wave solutions with nonzero components $v_2, T_{11}, T_{12}, T_{22}, T_{33}, \delta_{12}$ which depend on one spatial coordinate x_1 and time t.

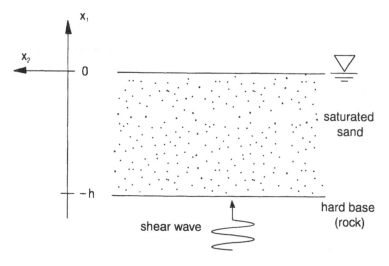

Figure 2: A layer of saturated sand.

The equation of motion for plane-wave solutions reduces to

$$\frac{\partial T_{12}}{\partial x_1} = \varrho\,\frac{\partial v_2}{\partial t}. \tag{6}$$

The constitutive equations are written as

$$\frac{\partial T_{11}}{\partial t} = H_{11}\left(T_{kl}, \frac{\partial v_2}{\partial x_1}, \delta_{12}\right), \tag{7}$$

$$\frac{\partial T_{12}}{\partial t} = H_{12}\left(T_{kl}, \frac{\partial v_2}{\partial x_1}, \delta_{12}\right), \tag{8}$$

$$\frac{\partial T_{22}}{\partial t} = H_{22}\left(T_{kl}, \frac{\partial v_2}{\partial x_1}, \delta_{12}\right), \tag{9}$$

$$\frac{\partial T_{33}}{\partial t} = H_{33}\left(T_{kl}, \frac{\partial v_2}{\partial x_1}, \delta_{12}\right), \tag{10}$$

$$\frac{\partial \delta_{12}}{\partial t} = F_{12}\left(\frac{\partial v_2}{\partial x_1}, \delta_{12}\right), \tag{11}$$

where T_{kl} denotes the components $T_{11}, T_{12}, T_{22}, T_{33}$. For purely shear waves the kinematic condition (4) is satisfied identically. Due to the volume conservation, the void ratio in each material point remains unchanged and, for simplicity, is taken to be uniform.

Let the motion of the layer be induced by a plane transverse wave coming from below. This is modelled by a periodic (sinusoidal) boundary condition for the velocity component v_2 at $x_1 = -h$. The upper boundary $x_1 = 0$ remains free of traction, which gives the boundary condition $T_{12} = 0$.

4 Dynamic solutions and liquefaction

The system (6)–(11) was solved by a finite difference technique. It should be mentioned that one faces considerable mathematical difficulties when solving boundary value problems for the materials which possess such properties as incremental non-linearity of the constitutive behaviour (this is the case for any plastic material) and pressure dependence of stiffness (this is the case for granular materials). These properties inevitably lead to the nonlinearity of the equations and thus raise the question of well-posedness of the problem. In most cases this question cannot be answered rigorously; however, in terms of classical solutions, the general reasoning suggests that such problem is most probably ill-posed, and therefore great caution should be exercised in the interpretation of numerical results. This issue and the numerical algorithm will be discussed in a companion paper.

Figures 3–5 show the solutions to the dynamic problem formulated above for a 20 m thick layer of saturated sand with a void ratio of 0.8. The frequency of the incoming wave is 5 Hz in all the cases. The figures show the distribution of the mean effective stress in the layer at different instances for three amplitudes of the boundary condition (velocity) v_0 at $x_1 = -20$ m. The dashed lines show the initial stress.

Disturbances of amplitudes smaller than 1 cm/s leave practically no permanent changes in the effective stress in the layer. This is a consequence of the fact that the behaviour of the soil is nearly elastic at small deformation. Disturbances of larger amplitudes

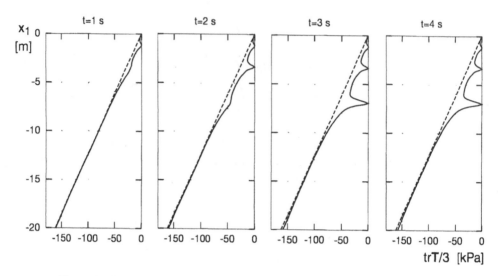

Figure 3: The mean effective stress in the layer at different instances. $v_0=2$ cm/s.

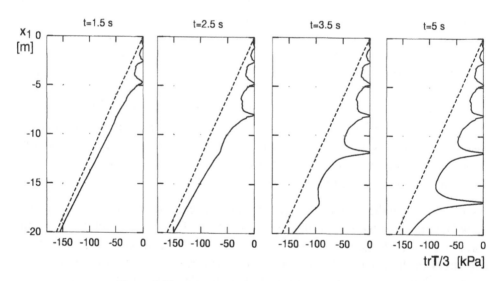

Figure 4: The same as in Figure 3 except for $v_0=4$ cm/s.

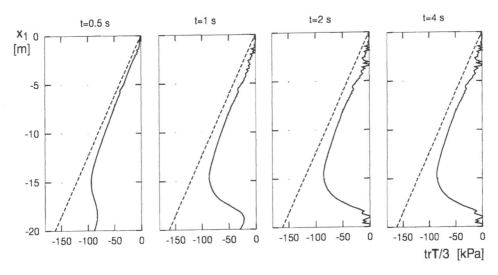

Figure 5: The same as in Figure 3 except for $v_0 = 12$ cm/s.

produce permanent changes in the effective stress which are found to be highly in-homogeneous.

At relatively small amplitudes, see Figure 3, the changes in the effective stress are observed only in the upper part of the layer. At larger amplitudes, Figure 4, the changes in the effective stress first appear in the upper part of the layer, like in the previous case, and then spread downwards. The most pronounced changes are observed in separate thin zones where the effective pressure is reduced to zero and the soil becomes fully liquefied. Thus, the liquefaction is localized in thin zones which emerge successively from the top to the bottom and divide the layer into several parts. If the amplitude is further increased, the average degree of liquefaction in the layer is increased as well: the effective pressure between the liquefaction zones may be reduced more than by half as compared to the initial value.

It is important to note that the location of the liquefaction zones in the cases shown in Figures 3,4 was found to depend on the numerical parameters and, in particular, on the degree of discretization. This fact suggests that the location of the liquefaction zones can hardly be predicted in reality, all the more because the parameters of the disturbace can be estimated only roughly. This is similar in a sense to the indeterminacy of the location of shear bands in the quasi-static problems of deformation of an initially homogeneous sample. The sensitivity of the location of the liquefaction zones to the numerical parameters appears to be attributed, at least in part, to the inhomogeneity of the initial stress and, specifically, to the low wave speed and small wave lengths in the vicinity of the free surface where the liquefaction originates. The solutions become robust and insensitive to the numerical parameters if the initial stresses in the layer are taken to be uniform so that the wave speed is initially uniform as well (the case which is, however, of limited practical interest).

If the amplitude of the incoming wave is rather large, as in Figure 5, the liquefaction pattern differs essentially from what has been considered so far. The liquefaction first occurs near the free surface like in the previous examples. However, right after that the soil becomes fully liquefied in the vicinity of the bottom. Since the liquefied soil has no shear stiffness, the motion of the base can no longer be transmitted to the layer, and the subsequent motion in the layer proceeds independently of the boundary condition. The motion of the soil rapidly decays in spite of the motion of the base. One can thus speak of the 'layer separation effect'. This effect also takes place in the cases shown in Figures 3,4 where each newly emerging liquefaction zone isolates the above situated soil from the lower part of the layer.

5   Settlement after liquefaction

Permanent changes in the pore water pressure give rise to the long-term process of seepage of the water and consolidation of the soil. We assume that the pore water pressure approaches its initial distribution and therefore so does the vertical effective stress. We are interested in the ultimate deformation of the soil rather than in the process of consolidation as such. The solution of the dynamic problem reveals that the vertical effective pressure is reduced so that the soil will eventually be compacted. The vertical deformation at a point x_1 can be found as the soluton of the problem of uniaxial compression of a dry soil and is expressed as

$$\varepsilon_{11}(x_1) = \int_{T_D(x_1)}^{T_0(x_1)} \frac{d\varepsilon_{11}}{dT_{11}} \, dT_{11}, \tag{12}$$

where $d\varepsilon_{11}/dT_{11}$ is the reciprocal of the stiffness at a given stress, $T_D(x_1)$ is the permanent vertical effective stress obtained from the solution of the dynamic problem, and $T_0(x_1)$ is the vertical effective stress that corresponds to the equilibrium distribution of the water pressure. The settlement of the layer and the change in the density can easily be calculated if the function $\varepsilon_{11}(x_1)$ is known.

Figure 6 shows the settlement of the free surface of a 20 m thick layer relative to the base as a function of the amplitude of the disturbance. Integral (12) was calculated with the same constitutive equation as in the dynamic problem. For comparison purposes the stress $T_D(x_1)$ was taken at times 2 s and 5 s, and the corresponding settlement is shown respectively in the left and right plots of Figure 6 . As was mentioned in Section 4, the location of the liquefaction zones is uncertain. In turn,

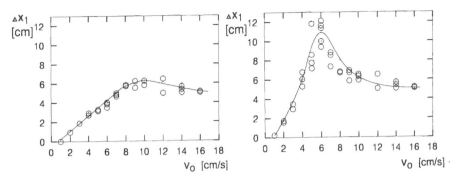

Figure 6: Settlement as a function of the amplitude. Duration of the disturbance: 2 s (left) and 5 s (right).

this leads to variations in the calculated settlement. This influence is indicated by the multiple points in Figure 6 which correspond to different numerical solutions.

Compaction and settlement of the layer begins with an amplitude of 1 cm/s and increases monotonically up to an amplitude of 9–10 cm/s for the 2 s duration and up to 5–6 cm/s for the 5 s duration. At larger amplitudes, the settlement decreases and then does not change any more. Another feature is that for large amplitudes the settlement after a 2 s disturbance is the same as after a 5 s disturbance.

The shape of the curves in Figure 6 is easily explained by the layer separation effect discussed above. Turning back to Figures 3,4, we see that the average reduction of the effective pressure in the whole layer is bigger in Figure 4 than in Figure 3 so that the resulting settlement is bigger as well. However, at strong disturbances, like in Figure 5, the most part of the soil has no time to liquefy as the bottom is rapidly liquefied and separates the layer; that is why the resulting settlement becomes smaller and why the curves in Figure 6 have a descending branch. Also seen from the Figure 5 is that, due to the same separation effect, there is no change in the effective stress after 2 s and therefore the settlement does not increase any more.

6   Acknowledgements

The support from the Deutsche Forschungsgemeinschaft (SFB 298 and SFB 461) is gratefully acknowledged. The authors are indebted to Prof. G. Gudehus for fruitful discussions.

References

[1] SEED, H. B. and LEE, K. L. (1966): *Liquefaction of saturated sands during cyclic loading.* J. Soil Mech. Found. Div., Proc. ASCE, 92, SM6, 105–134.

[2] PEACOCK, W. H. and SEED, H. B. (1968): *Sand liquefaction under cyclic loading simple shear conditions.* J. Soil Mech. Found. Div., Proc. ASCE, 94, SM3, 689–708.

[3] KOLYMBAS, D. and WU, W. (1993): *Introduction to hypoplasticity.* In: *Modern Approaches to Plasticity*, D. Kolymbas (ed.), Elsevier, Amsterdam, 213–223.

[4] GUDEHUS, G. (1996): *A comprehensive constitutive equation for granular materials. Soils and Foundations*, 36(1), 1–12.

[5] VON WOLFFERSDORFF, P. A. (1996): *A hypoplastic relation for granular materials with a predefined limit state surface. Mechanics of Cohesive-frictional Materials*, 1(3), 251–271.

[6] NIEMUNIS, A. and HERLE, I. (1997): *Hypoplastic model for cohesionless soils with elastic strain range. Mechanics of Cohesive-frictional Materials*, 2(4), 279–299.

[7] BAUER, E. (1996): *Calibration of a comprehensive hypoplastic model for granular materials. Soils and Foundations*, 36(1), 13-26.

[8] HERLE, I. and GUDEHUS, G. (1999): *Determination of parameters of a hypoplastic constitutive model from properties of grain assemblies. Mechanics of Cohesive-frictional Materials*, 4(5), 461-486.

Numerical Analysis of a Dynamic Intensive Compaction

Konrad Nübel, Andreas Schünemann, Sven Augustin

Institute of Soil and Rock Mechanics, Technical University of Karlsruhe, Germany

Abstract: In order to estimate the densification effect of a dynamic intensive compaction on a slight cohesive granular material a calculation procedure is proposed. An equation of motion which can be solved by a standard numerical procedure is established to calculate the penetration depth of the compaction weight into the ground. With a given compaction raster and the calculated depth of influence the decrease of the average void ratio can be calculated for the compacted zone. With an one-dimensional calculation and the hypoplastic constitutive law settlements of a given load can be estimated before and after the compaction.

1 Introduction

The dynamic intensive compaction (Fig. 1) is a common method to densify slightly cohesive granular material in order to reduce settlements for a new building. For the planning of a project the dimensions of the dynamic intensive compaction (mass-, dimension-, drop down height- of the falling weight, dimension of compaction raster, number of compactions for one location) should be designed for the required maximum settlement of the specific construction. It is almost impossible to design the dimensions with conventional soil mechanics. The density of the compacted layer and consequently the stiffness changes dramatically in the process of compaction. However, the stiffness of the soil in conventional soil mechanics is a material parameter which has to be determined experimentally for any potential density. This is rather uneconomical and inaccurate.

In the hypoplastic constitutive law (e.g. [7], [4], [12]) the stiffness is defined by the ratio of stress increment and strain increment dependent on the actual state of the material. The actual state in the constitutive model is totally determined by the stress tensor $\mathbf{T}$ and the void ratio e. Knowing the void ratio and the stress state before and after the compaction it is possible to estimate the reduction of settlement of a given load. In the following the applied calculation procedure is described.

2 Material and Model Parameters

The granular soil in the hypoplastic model can be described by 8 material parameters, which can be easily determined by granulometric properties and standard tests

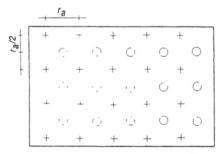

+ primary compaction point

○ secondary compaction point

Figure 1: Dynamic intensive
compaction.

Figure 2: Typical compaction raster.

on disturbed samples (s. HERLE and GUDEHUS [6]). The parameters are the critical friction angle φ_c, the granular hardness h_s, the exponent of compression n, the minimum, maximum and critical void ratios e_{d0}, e_{c0}, e_{i0} at pressure zero and the exponents α and β.

For the nonlinear calculation the accurate determination of the initial state is of importance. The effective vertical and horizontal stresses are commonly calculated as

$$\sigma'_v = \gamma h \qquad \text{and} \tag{1}$$

$$\sigma'_h = (1 - \sin \varphi_c)\sigma'_v . \tag{2}$$

Alternatively the horizontal stress σ_h can be obtained by means of field measurements (CUDMANI [2]). The in situ void ratio e_{por0} should be measured from undisturbed samples or by standard or vibro penetration in connection with the evaluation method proposed by CUDMANI [3] and related to pressure zero.

Further model parameters for the compaction and settlement calculation are listed in Tab. 1

3 Penetration depth of the falling weight

The behaviour of the unsaturated soil at the beginning of the penetration resembles a drained deformation path. With increasing penetration depth the pore pressure

Table 1: Parameters for the compaction and the load for the settlement calculation

value	symbol	unit
falling weight		
edge length	b	[m]
drop down height	h	[m]
mass	m_{Pl}	[t]
raster length	r_a	[m]
in situ state of the soil		
void ratio	e_{por0}	[-]
capillary choesion	c_c	[kPa]
unit weight	γ	[kN/m^3]
load on the ground		
length	a_l	[m]
width	b_l	[m]
pressure	p_l	[kPa]

increases and the soil exhibits an undrained response. The undrained cohesion can be calculated by the hypoplastic model as follows [5]

$$c_u = p_s \frac{3\sqrt{6}}{4a}(f_d - \sqrt{f_d^2 - 1}), \tag{3}$$

where $p_s = -1/3\,(\sigma_v + 2\sigma_h)$ is the initial isotropic pressure. Thus the increase of the undrained cohesion with the depth is taken into consideration. The scalar factor f_d is defined as

$$f_d = \left(\frac{e - e_d}{e_c - e_d}\right)^\alpha, \tag{4}$$

where $f_d \geq 1$ for the loose soil. The Parameter a is depending on the critical friction angle, i.e.

$$a = \frac{\sqrt{3}(3 - \sin \varphi_c)}{2\sqrt{2} \sin \varphi_c}. \tag{5}$$

The dependence of the undrained cohesion on the shear velocity can be considered by the equation given by LEINENKUGEL [8], i.e.

$$c_u \approx c_{u_\alpha}[1 + I_{v_\alpha} \ln(\dot{\gamma}/\dot{\gamma}_\alpha)]. \tag{6}$$

The viscosity index I_{v_α} with a common value from 0.5 to 3.0% increases with decreasing main grain size d_{50}. The falling weight penetrates the ground at $t = 0$ with high velocity:

$$v_0 = \sqrt{2gh}. \tag{7}$$

The base failure F_0 is related to the friction angle and the shape of the falling weight according to:

$$F_0 = b^2 \left[\gamma\, b\, N_b \cdot 0,7 + \gamma\, s_0\, N_q (1 + \sin \varphi_c) \right] \; . \tag{8}$$

Following GUDEHUS [9] N_b and N_q can be calculated as follows:

$$N_b = \frac{1}{2} \exp \left[5,71 (\tan \varphi_c)^{1,15} \right] \; , \tag{9}$$

$$N_q = \frac{1 + \sin \varphi_c}{1 - \sin \varphi_c} \exp \left(\pi \tan \varphi_c \right) \; . \tag{10}$$

The settlement s_0 of the falling weight up to the base failure can be calculated according to chapter 4. The driving force to achieve the base failure increases with the depth of penetration s and can be approximated by the function $F = F_0 (s/s_0)^\kappa$ (GUDEHUS [9]) where the value $F_1(s_1)$ has to be calculated in a depth far away from the surface by

$$F_1 = \left[(2 + \pi) c_u + \gamma s_1 \right] b^2 \; , \tag{11}$$

yielding the exponent κ as (s. Fig. 3)

$$\kappa = \frac{\ln \left(F_1 / F_0 \right)}{\ln \left(s_0 / s_1 \right)} \; . \tag{12}$$

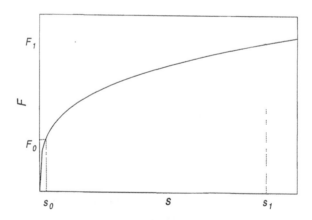

Figure 3: Evolution of the penetration resistance F

Assuming a plastic impact the resulting velocity of the masses is

$$v_{res} = \frac{m_1 v_1 + m_2 v_2}{m_1 + m_2} = \frac{m_{Pl}\, v_{Pl} + 0}{m_{Pl} + m_{Bo}} \; . \tag{13}$$

The total mass of the accelerated body can be approximated as the sum of the falling weight m_{pl} and the mass of a pyramid under the plate m_{Bo}, i.e.

$$m_{Pl} + m_{Bo} = m_{Pl} + \frac{\gamma b^3}{3g} . \tag{14}$$

The resulting equation of motion which is a nonlinear ordinary differential equation of second order can be transformed in an ordinary differential equation of first order, yielding

$$\frac{ds}{dt} = \sqrt{2}\sqrt{g\,s - \frac{C_m}{\kappa+1} s^{\kappa+1} + \frac{V_0^2}{2}} , \tag{15}$$

with

$$C_m = \frac{F}{m_{Pl} + m_{Bo}} \left(\frac{1}{s_0}\right)^\kappa . \tag{16}$$

Consequently, the numerical solution by RUNGE and KUTTA [11] can be applied to solve eq. (15). The numerical procedure is truncated as soon as the penetration resistance F reaches a plateau. Therefore a rebound of the mass is neglected assuming a pure plastic deformation of the loose soil. The penetration depth after the first penetration cycle is then

$$s_{c1} = s(t)_{1,\text{max}} . \tag{17}$$

The calculation of the penetration depth for further compaction cycles can be repeated in a similar manner. The thickness of the compacted layer which volume change in the further compaction cycles can be neglected, can be approximated as $d_{d1} = 1 \cdot b$, since the mechanism of penetration is a compaction with following replacement. The increased base resistance can be obtained by eq. (8) and (11) with the changed penetration depth $d = s_{c1}$. With the knowledge of these values the calculation procedure can be carried on in the same manner. The total penetration depth and the thickness of the compacted layer t_{total} approach limit values after a particular number of compaction cycles. A reasonable assumption thereby is that the limit value of t_{total} is reached if the penetration depth s_n of a compaction cycle is lower than 15% the edge length b of the falling weight.

4   Calculation of settlements

The reference settlement of a given load before and after the compaction is calculated in the centre of the load. Following [10] the settlements can be calculated using the hypoplastic constitutive law by subdividing the soil column under the center of the load and simulating a triaxial compression for each sublayer. For the triaxial compression the ratio of radial and horizontal strain increment $\vartheta = \dot{\varepsilon}_r / \dot{\varepsilon}_a$ can be approximated as follows: With

$$\alpha = \frac{b_l}{a_l} \frac{\pi}{2}; \quad [\text{rad}] \quad a_l \geq b_l \tag{18}$$

and

$$\vartheta_{\infty,R} = -\sqrt{(0,2 \cos\alpha)^2 + (0,3 \sin\alpha)^2}, \tag{19}$$

the ratio ϑ can be expressed as

$$\vartheta(z,b) = \begin{cases} \vartheta_{\infty,R} \cdot \dfrac{|z|}{2\,b_l} & \text{für} \quad \dfrac{|z|}{2\,b_l} \leq 1, \\[3mm] \vartheta_{\infty,R} & \text{für} \quad \dfrac{|z|}{2\,b_l} > 1. \end{cases} \tag{20}$$

Consequently the triaxial compression is totally determined by the boundary conditions and can be solved sufficiently accurate by a EULER forward integration scheme using the following equations of the hypoplastic model for triaxial compression [5]:

$$\dot{\varepsilon}_a = \dot{\sigma}_a \left\{ f_b\, f_e\, \frac{(\sigma_a + 2\sigma_r)^2}{\sigma_a^2 + 2\sigma_r^2} \left[1 + a^2 \frac{(\sigma_a + 2\vartheta\sigma_r)\sigma_a}{(\sigma_a + 2\sigma_r)^2} - f_d \frac{a}{3} \frac{5\sigma_a - 2\sigma_r}{\sigma_a + 2\sigma_r} \sqrt{1 + 2\vartheta^2} \right] \right\}^{-1}, \tag{21}$$

$$\dot{\sigma}_r = f_b\, f_e\, \frac{(\sigma_a + 2\sigma_r)^2}{\sigma_a^2 + 2\sigma_r^2} \left[\vartheta\dot{\varepsilon}_a + a^2 \frac{(\sigma_a + 2\vartheta\sigma_r)\dot{\varepsilon}_a\sigma_r}{(\sigma_a + 2\sigma_r)^2} + f_d \frac{a}{3} \frac{4\sigma_r - \sigma_a}{\sigma_a + 2\sigma_r} \sqrt{\dot{\varepsilon}_a^2(1 + 2\vartheta^2)} \right]. \tag{22}$$

The evolution equation of the void ratio thereby, is:

$$\dot{e} = (1 + e)(\dot{\varepsilon}_a + 2\dot{\varepsilon}_r). \tag{23}$$

By the scalar factors f_d and f_e the dependency of the void ratio and the isotropic pressure is considered (BAUER and Wu [1]). Factor f_d is defined via eq. (4) and factor f_e as

$$f_e = \left(\frac{e_c}{e}\right)^\beta. \tag{24}$$

Factor f_b is the consistency factor in the model (GUDEHUS [4]) and can be expressed as

$$f_b = \frac{h_s}{n} \left(\frac{e_{i0}}{e_{c0}}\right)^\beta \frac{1 + e_i}{e_i} \left(\frac{3p_s}{h_s}\right)^{1-n} \cdot \left[3 + a^2 - a\sqrt{3} \left(\frac{e_{i0} - e_{d0}}{e_{c0} - e_{d0}}\right)^\alpha\right]^{-1}. \tag{25}$$

The additional vertical stress $\Delta\sigma_v$ imposed by the load can be calculated after STEIN-BRENNER as

$$\Delta\sigma_a = 4\,f_\sigma\,p_l\;,\tag{26}$$

$$f_\sigma = \frac{1}{2\pi}\left[\arcsin\frac{a_l b_l}{\sqrt{a_l^2+z^2}\,\sqrt{b_l^2+z^2}} + \frac{a_l b_l |z|(r^2+z^2)}{r(r^2 z^2 + a_l^2 b_l^2)}\right]\;,\tag{27}$$

with

$$r = \sqrt{a_l^2 + b_l^2 + z^2}\;.\tag{28}$$

A typical compaction raster is shown in Fig. 2. The compacted volume of a single square of the raster is

$$V_d = r_a^2\,t_{\text{total}} - 2b^2 s_{\max}\;,\tag{29}$$

where r_a is the raster length, t_{total} the calculated thickness of the compacted layer, b the edge length of the falling weight and $s_{\max}$ the total penetration depth as mentioned before. The decreased void ratio e_{por0}^{vd} of the compacted layer can then be approximated by

$$e_{\text{por0}}^{vd} = \frac{r_a^2\,t_{\text{total}}}{V_d}(1 + e_{\text{por0}}) - 1\;.\tag{30}$$

The horizontal pressure of the compacted layer increases during the compaction cycles and therefore the stiffness. However, the value of the horizontal stress is not known after the process of densification. Therefore it is convenient to assume a lower value of $0.5 \cdot \vartheta_{\infty,R}$ in eq. (19).

Table 2: Hypoplastic parameters of the open mine fill material

φ_c [°]	h_s [MPa]	n	e_{d0}	e_{c0}	e_{i0}	α	β
30.0	80.0	0.24	0.61	1.10	1.27	0.10	2.40

5 Behaviour of the Model and Verification

The proposed calculation procedure was implemented in an C++ code. In order to verify the numerical analysis a dynamic intensive compaction of an open mine fill was analysed in detail. The material parameter and the in situ state of the slight cohesive granular material were determined (s. Tab. 2). The penetration depth of the falling weight was measured after each cycle. The thickness of the compacted layer was monitored by means of a tube which was placed into the ground in the middle of a compaction point. The tube was able to reduce its length almost without any resistance force. A monitoring device which was brought into the tube measured the

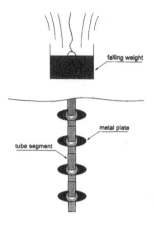

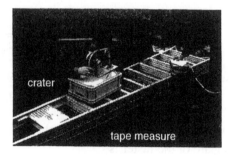

Figure 4: Monitoring device.   Figure 5: Measurement in the crater.

position of metal plates in the ground which were placed around the tube in certain levels (Fig. 4, 5). The results of the measurements are shown in Fig. 6.

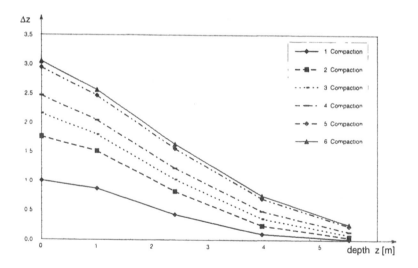

Figure 6: Measurement of the impact of dynamic densification.

All relevant system parameters are listed in Tab. 3. In Fig 7 the accumulated penetration depth of the numerical analysis is presented. With increasing number of compaction cycles the incremental penetration depth decreases. The total penetration depth after six cycles is slightly higher than in the field test (Fig 6). The decrease of the maximum settlement in the middle of a given load can be seen in Tab. 4.

Table 3: Parameters for the compaction and the load for the settlement calculation

b [m]	h [m]	m_{Pl} [t]	r_a [m]	e_{por0} [-]
1.7	18.5	15.5	6.0	1.14
c_c [kPa]	γ [kN/m^3]	a_l [m]	b_l [m]	p_l [kPa]
10.0	17.0	10	10	100

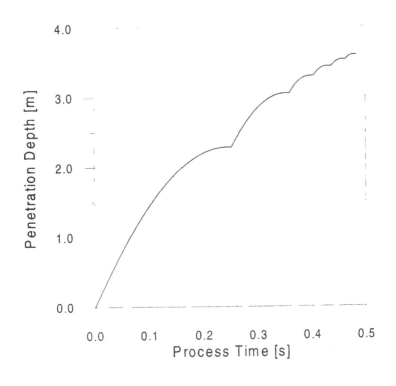

Figure 7: Calculated penetration depth after six compaction cycles.

Table 4: Parameters for the compaction and the load for the settlement calculation

penetration depth	d_{max}	3.59	[m]
initial void ratio	e_{por0}	1.14	[-]
void ratio after compaction	e_{por0}^d	0.92	[-]
thickness of compacted layer	t_{total}	5.68	[m]
initial settlement	s_0	-40.7	[cm]
settlement after compaction	s_1	-34.0	[cm]

6 Conclusion

The comparison between the proposed numerical analysis and the field measurements shows a good accordance. The penetration depth in the prediction of the model is somewhat too high, the thickness of the compacted zone is in good agreement with the measurements. However, it should be noted that the behaviour in situ is quite sensitive against small fluctuations of state and material parameters considering the relative small size of the falling weight. Therefore the calculated penetration depth should rather be used as an approximation than an exact prediction.

References

[1] BAUER, E. and WU, W. (1994): *Extensions of a Hypoplastic Constitutive Model with Respect to Cohesive Powders.* Proc. of the 8th Int. Conference on Computer Methods and Advances in Geomechanics, Morgantown, West Virginia (USA), 513–536.

[2] CUDMANI, R. (1996): *Anwendung der Hypoplastizität zur Interpretation von Drucksondierwiderständen in nichtbindigen Böden.* Geotechnik, 19(4),266–273.

[3] CUDMANI, R. and HUBER, G. (2000): *Evaluation of the State of Cohesionless Open Mine Fills Before and After Blasting Compaction by Means of Field Testing.* To be published in: Internationaler Workshop on Compaction of Soils, Granulates and Powders, Innsbruck, Austria.

[4] GUDEHUS, G. (1996): *A comprehensive equation for granular materials.* Soils and Foundations, 36(1):1–12.

[5] HERLE, I. (1997): *Hypoplastizität und Granulometrie einfacher Korngerüste.* Veröffentlichung des Institutes für Bodenmechanik und Felsmechanik der Universität Fridericiana in Karlsruhe, Heft 142.

[6] HERLE, I. and GUDEHUS, G. (1999): *Determination of Parameters of a Hypoplastic Constitutive Model from Properties of Grain Assemblies* Mechanics of Cohesive-Frictional Materials, 4(5):461–486.

[7] KOLYMBAS, D. (1999): *An Outline on Hypoplasticity.* Archive of Applied Mechanics, 61:143–151.

[8] LEINENKUGEL, H.J. (1976): *Deformations- und Festigkeitsverhalten bindiger Erdstoffe.* Veröffentlichung des Institutes für Bodenmechanik und Felsmechanik der Universität Fridericiana in Karlsruhe, Heft 66.

[9] MELHORN G. and GUDEHUS, G. (1995): *Geotechnik* Der Ingenieurbau, Ernst & Sohn.

[10] NÜBEL, K., KARCHER, C., HERLE, I. (1999): *Ein einfaches Konzept zur Abschätzung von Setzungen.* Geotechnik, 22(4):251–258.

[11] PRESS, W.H., TEUKOLOVSKY, S.A., VETTERLING, W.T., FLANNERY, B.P. (1992): *Numerical Recipies.* Cambrige University Press, Second Edition.

[12] VON WOLFFERSDORFF, P.-A. (1996): *A hypoplastic relation for granular materials with a predefined limit state surface.* Mechanics of Cohesive-Frictional Materials, 1:251–271.

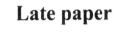

Late paper

Experimental investigation of blasting compaction of a cohesionless pit mine fill

Gerhard Huber and Roberto Cudmani

Institute of Soil Mechanics and Rock Mechanics, University of Karlsruhe, Germany

Abstract: The mechanical behaviour of a very loose cohesionless open pit mine fill during and after single and group charge blasting compaction is experimentally investigated. Considering the distance from the charges the parts of the fill affected by single or group charge blastings can be classified into near, intermediate and far fields. In the near field the rapid cavity expansion originated by the blasting induces a shock front whose propagation velocity decreases with the distance from the source resulting in obliteration of the grain skeleton. Soil compaction takes place mainly in the intermediate field. The compression of the soil skeleton left behind by the shock wave leads to a rapid increase of the effective pressure followed by a delayed increase of pore water pressure due to the presence of gas bubbles. The loss of effective pressure (liquefaction) starts a short time after detonation and reaches distances of about 20 m, leading to a collapse of the grain skeleton and hence to a compaction zone within the fill. Simultaneous measurements of earth and pore water pressures enabled the detection of liquefaction and its duration.

1 Introduction

The Lausitz region in the eastern part of Germany is a major lignite reservoir. Former procedures of refilling open pit mines resulted in the formation of very loose cohesionless deposits with a strong tendency towards spontaneous liquefaction after flooding. Remediation is required to open these waste lands for public use, but also due to the latent danger the current situation represents for people ([3], [7]).

Considering the large areas involved, an economical method to stabilize the fills under groundwater table is blasting compaction ([2], [6]). However, since the mechanisms underlying densification of very loose granular soils due to blasting have not yet been completely understood the use of this method in practice is based mostly on empirical rules.

2 Instrumentation

In order to investigate the mechanical behaviour of loose cohesionless deposits during various stages of single and group blasting compaction two field test series were carried out in the vicinity of the Koschen Lake in the state of Brandenburg. In both cases the test fields were extensively instrumented in order to measure shock wave velocity, particle velocity both in the ground and on the surface, surface settlements, total horizontal earth pressures and pore water pressures. Short and long time responses of the fill were accurately monitored.

2.1 Earth and pore water pressure

In order to monitor the evolution of effective pressure during and after blasting total (earth) and pore water pressure were simultaniously measured using combined push-in spade shaped gauges (C-gauges). Additional measurements of total pressure and pore water pressure were performed separately using earth pressure gauges (E-gauges) and pore water pressure (P-gauges), respectively.

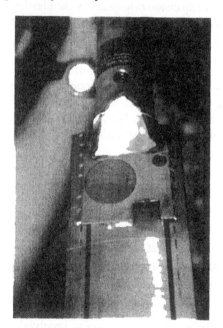

Figure 1: Earth pressure gauges: E-gauge (left), C-gauges (right).

The earth pressure gauge is a spade-shaped fluid filled pressure transmitter connected to a pressure transducer built into a mounting head. Conventional earth pressure gauges use a hydraulic/pneumatic compensation system with a response time of about 10 – 20 s. However, for measurements during blastings gauges equipped with pressure transducers with a response time of 5 - 30 ms were required. The electrical output signal of the pressure transducer is proportional to the pressure acting on the surface of the membrane plates. The response of gauge was tested in a water filled pressure chamber and compared to the response signal of a hydrophone (bandwidth 150 kHz). The delay and rise of the pressure gauge response was shorter than 10 ms, this is equivalent to a bandwidth ≥ 100 Hz and sufficient to accurately measure the expected dynamical changes of earth pressure.

The pore water pressure gauges consist of a pressure chamber with a stiff filter plate connected to a pressure transducer. In the calibration chamber the gauges showed nearly the same bandwith as the earth pressure gauges. However, the bandwidth can be reduce down to 10 Hz if the filter plate is not saturated or if the chamber is not completely fluid filled.

The installation of C- and E-gauges up to a depth of about 17 m was performed in four steps: drilling a fluid stabilized borehole up to 1 m above the final depth, pushing in the gauge to the final depth, decoupling the rod from the mounting head of the gauge (necessary since the pressure transmitter is sensitive to bending) and filling the borehole with gravel. The P-gauges were pushed in into the desired depths using the equipment for cone penetration tests.

Installing the gauges caused unavoidably soil disturbance up to a certain extent. A numerical investigation of the cavity expansion problem in cohesionless soils shows that the zone affected by the cavity expansion increases with the density of the soil ([4]). Therefore, for a loose soil only a small disturbance is expected. Preliminary tests in the laboratory confirmed the results of the numerical simulation.

Ten days after the installation the measured earth pressure reached almost constant values. The porewater pressure gauges showed unchanged values already some hours after the installation.

2.2 Particle velocity and acceleration

The geophones and the accelerometers were built in casings with approximately the same density as the expected surrounding soil in order to minimize the influence of inertial forces on the measured signals. They were also pushed in using the equipment of the cone penetration tests and could therefore not be retrieved after the tests.

1-D and 3-D geophones were built using geophone base units with a natural frequency of 8 Hz and an upper limiting frequency of about 200 Hz (Figure 2).

In order to determine the decrease of the shock wave velocity in the vicinity of the blasting point B five accelerometers were placed at different distances from the blasting point at the same depth as the charge. The accelerometers were expected to be destroyed by the shock wave. The accelerometer had a bandwidth going from 1 Hz to 3 kHz.

Figure 2: 1D- and 3D-geophones: mounting frame (top), epoxy resine casing (bottom)

3 Single point blasting

The site chosen was located in the area of Kleinkoschen near Senftenberg in the eastern part of Germany (for further details see [1]). The deposit consists of three layers with a total thickness of 42 m. The two top layers (0 – 7, 9 – 16 m) consist of very loose fine and medium quartz sands with tip restistances q_c between 2 – 3 MPa. The third layer has some content of fine cohesive material. The layers are separated by working planes (1 – 2 m), which were compacted by heavy machinery during refilling. The ground watertable was at a depth of about 5 m below ground level. The ground level was at 103 m NN.

The instrumentation of the test field began with the installation of the earth pressure gauges two months before blasting and ended with the placement of the accelerometers some days before blasting.

The investigation of single point blasting included three blasting points (A, B, C) (Figure 3).

Each blasting point consisted of three charges (C1, C2, C3) placed at depths of 7, 15 and 25 m. The charges C1, C2 and C3 were 2.5, 5.0 and 7.5 kg, respectively (Figure 4). The three charges were connected to separated detonation cords and detonated with a time delay of about 6 ms in the sequence C1-C2-C3 by means of electrical detonators.

3.1 Results

Settlements were recorded up to a distance of about 30 m from the blasting points. Maximum settlement of about 1.5 m was measured near the blasting points. Cracks were observed on the ground surface at various distances from the blasting points. The crack width varied between 5 and 20 mm and their depths were larger than 1 m.

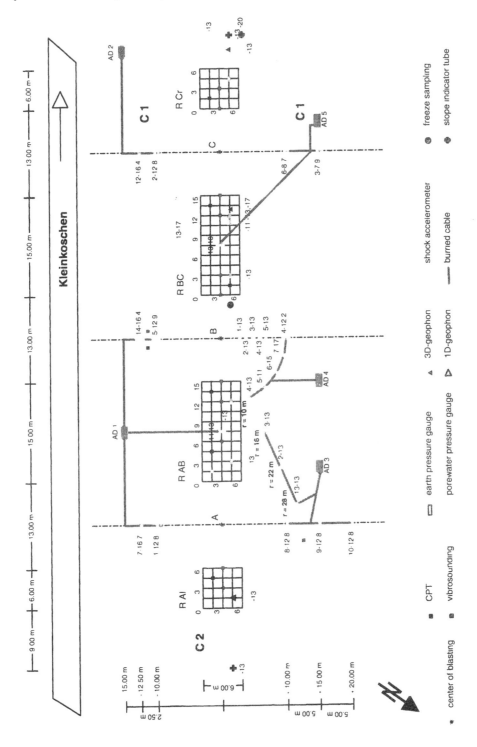

Figure 3: Field test: instrumentation single point blasting (A, B and C)

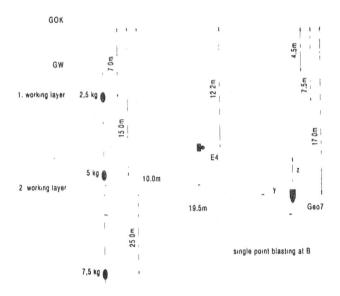

Figure 4: Single point blasting at B: placement of charges, geophone Geo7 and earth pressure gauge E4.

In the vicinity of blasting point B the initial velocities of the shock front at a depth of 15 m (see also figure 4) were determined using the accelerometer signals:

distance 3 – 5 m   1094 m/s,
distance 3 – 6 m   1033 m/s,
distance 3 – 7 m   979 m/s.

Three of the five transducers were destroyed during blasting. Similar shock front velocities were also determined by Raju [6] by placing the transducers at various depths above the charge.

Figure 5 shows the evolution of the effective pressure (p_{ef}) at different distances from blasting point B based on the evolution of total pressure (p_{tot}) and pore water pressure (p_w) measured at the C-gauges E1, E2 and E3. The shock front induces a steep increase followed by a slower decrease of p_{ef}. The strongest increase and decrease of p_{ef} was obtained at E4 since this was the gauge nearest to B. However, though the decrease of p_{ef} at E4 was considerable a complete loss of effective pressure (liquefaction) did not take place at the Loacation of gauge E4. All gauges showed a residual increase of p_{ef} (5 – 22 kPa) after blasting.

Figure 6 shows the evolution of pore water pressure at different depths and distances from point B. Though all gauges show an increase of p_w the shape of the curves differs among the gauges. Figure 7 shows the pore water response of all gauges placed at a distance of 10 m. The steepest increase is shown by P13 with a peak rise

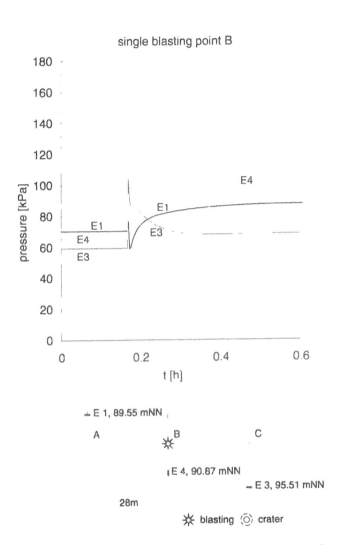

Figure 5: Evolution of p_{ef} after single point blasting at B

time of 25 s followed by P7 with 55 s. For comparison at gauge P6 the maximum
value of p_w was reached only after 450 s.

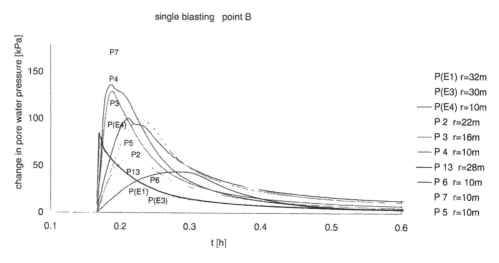

Figure 6: Evolution of p_w after single point blasting at B

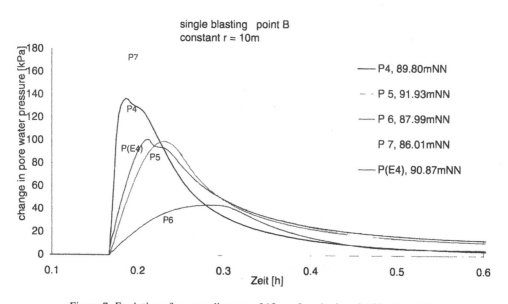

Figure 7: Evolution of p_w at a distance of 10 m after single point blasting at B

Figure 8 compares the evolution of particle velocity, total pressure and pore water
pressure during the first second after blasting. The electrical trigger signal used to
ignite the detonator is also shown in the figure. Figure 4 shows the positions of the

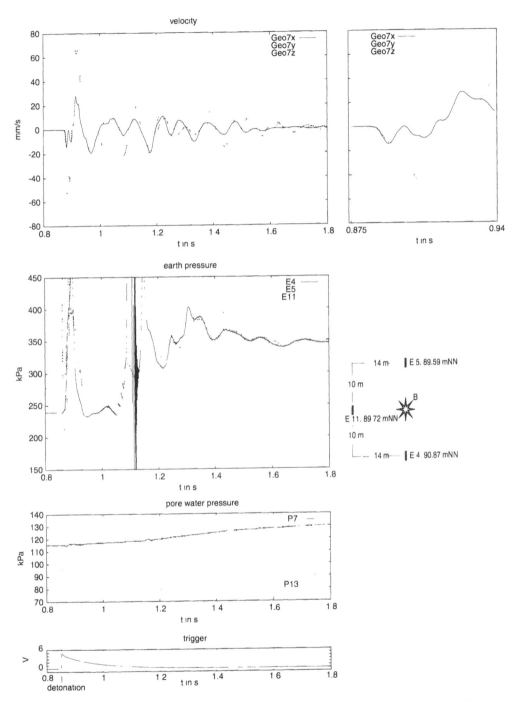

Figure 8: Single point blasting at B: (from top to bottom) particle velocity at Geo7, total pressure at E4, E5 and E11, pore water pressure at P7 and P13 and detonation trigger signal

gauges E4 and Geo7 with respect to the three charges.

Following the sequence of detonation the radial component of the particle velocity (Geo7y) decreases in three steps reaching a velocity of -68 mm/s in 30 ms. Since the charge C2 is the nearest to the geophone the front induced by this charge arrives at first, followed by the front induced by the detonation of charges C1 and C3, respectively. Detonation of C3 produced a reversal of the vertical component of particle velocity (Geo7z) since C1 and C3 were placed above and beneath of the geophone. The magnitude of the vertical component was almost the same as that of the radial component. On the other hand, according to our expectation the tangential component of particle velocity (Geo7x) showed the smallest magnitude among the three components (theoretically, Geo7x should vanish if a pure spherical cavity occurred). Integration of radial component of particle velocity in the first 30 ms leads to a maximal radial displacement of about 2 mm.

A short time after blasting the total pressure p_{tot} increases rapidly, reaching a peak after $t \approx 50$ ms and decreasing to nearly its initial value. After $t \approx 150$ ms p_{tot} begins to increase again reaching after further 800 ms an almost stationary value. Approximately the same increase of total pressure ($\Delta p_{tot} \approx 100$ kPa) was measured by the three gauges. On the other hand, the gauge P7 positioned at approximately the same depth and distance from B as the earth pressure gauges shows in the same time scale a weak and almost continuous increase of p_w of ≈ 15 kPa. Qualitatively the same behaviour is shown by gauge P13 at a larger distance from B than P7.

The evolution of earth pressure and pore water pressure indicates that plastic deformation of the soil takes place some time after the passage of the first wave front. The reason for the observed increase of effective pressure seems to be the compression caused by a "compaction wave" coming after the shock front. The fact that the compression of the grain skeleton does not lead to a fast increase of pore water pressure seems to be a consequence of the presence of gas bubbles. In fact if the degree of saturation of the soil were nearly to 100% a much faster rise of pore water pressure would be expected. On the contrary, if the soil contains gas bubbles the initial compression leads to a decrease of the volume of gas bubbles and to a small increase of pore water pressure. The presence of gas bubbles was confirmed by freeze sampling. The measured degree of saturation was between 80 to 95% ([5]).

Figure 10 shows the evolution of the total pressure, pore water pressure and effective pressure at gauge E2 after the second blasting test (point C). It can be seen that p_{ef} vanished approximately 45 s after blasting indicating that the soil at this location liquefies. After the liquefaction phase, which has a duration of about two minutes, p_{ef} increases again and reaches a stationary value which is lower than the initial p_{ef} (see figure 9). In this case it seems to be impossible to detect liquefaction and determine its duration by monitoring pore water pressure only.

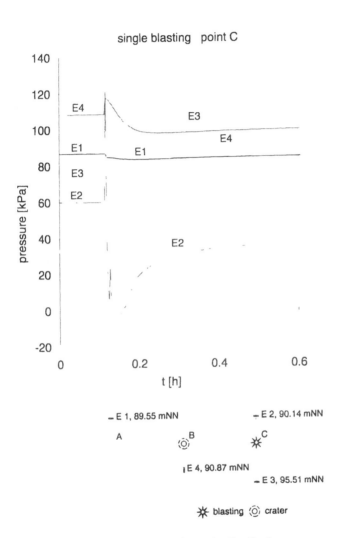

Figure 9: Earth pressure single blasting, point C: effective pressure $p_{tot} - p_w$

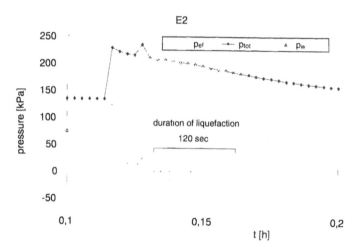

Figure 10: Liquefaction during blasting, point C

In the first seconds after detonation strong vibrations were observed on the ground surface in near and intermediate fields. The rate of settlement increases from the very beginning, achieves a maximum in the phase of liquefaction and vanish only after a few days. The evolution of settlements shows that compaction mainly occurs during and after the liquefaction phase.

Finally, the evolution of p_{ef} obtained at different C-gauges in the first 30 minutes after the second blasting test (point C) is compared in Figure 9. As shown no liquefaction could be detected at gauges E1, E3 and E4. At E1 and E4 the stationary values of p_{ef} are lower compared to their initial values. Therefore, the increase of p_{ef} induced at E1 and E4 by the first blasting test (point B) was partially lost after the second blasting test (point C). On the contrary, at E3 an increase of the stationary effective pressure took place since the soil at E3 was almost not affected by the first blasting test.

4   Blasting in groups

Based on the results of the single charge blasting tests four different group charge blastings (V1, V2, V3, V4) were performed to investigate the influence of charge arrangement on the effectiveness of compaction. Each group consisted of four blasting points (A1, A2, B1, B2) arranged in a rombical shape. Each blasting point was composed of three charges with the same masses as in the case of single point blasting.

The lenght of the short axes of the rombus (A1-A2) and the distance between adjacent blasting points were 20 m. The region within the rombus was expected to liquefy since in the single point blasting soil liquefaction had extended up to a distance of about 10 m from the blasting points.

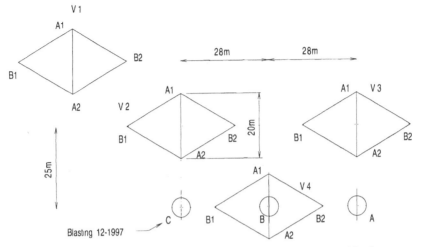

Figure 11: Topview of the field test used in the investigation of group charge blasting compaction.

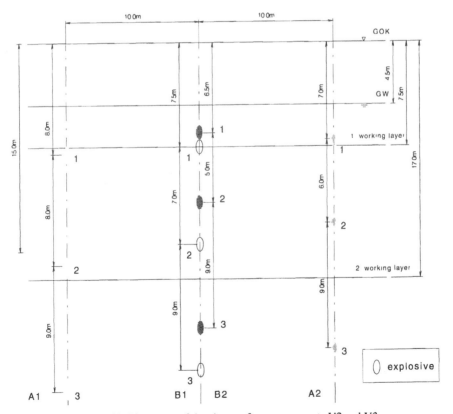

Figure 12: Placement of the charges for arrangements V2 and V3.

Figure 11 shows the topview of the field test used in the investigation. In rombus V1 the charges were placed at the same depth as for single point blasting and detonated in the sequence A2-B2-B1-A1 with time delays of 0, 2, 4, 6 s. This is one of the arrangements commonly used for blasting compaction. In rombus V2 the charges were placed at the depths shown in figure 12. They detonated in the sequence A1-A2-B1-B2 with time delays 0, 2, 4, 6 s.

In rombus V3 the same charge arrangement and detonation sequence as in V2 was used. The time delays were 0, 6, 66, 72 s. In rombus V4 the effect of re-compaction was investigated but this arrangement will not be further considered.

The evolution of settlement, pore water pressure and total pressure were similar to those measured in the single point blasting tests. No significant difference of compaction between the "conventional" arrangement (V1) and the arrangements V2 and V3 was shown neither by field testing [1] nor by surface levelling. The lowest level of the vibration in the intermediate and far field was induced by the blasting of arrangement V3 since the duration of the vibrations produced by the single detonations (A1, A2, B1 and B2) was smaller than the shortest detonation delay utilized.

5   Conclusions

The process of blasting compaction of a very loose cohesionless fill from open pit mines was experimentally investigated. The mechanical behaviour during and after blasting compaction was monitored by measuring particle velocities and accelerations, total pressure and pore water presure. Two types of blasting tests were performed: single point blasting with three charges and group charge blasting with delayed detonations. According to the distance from the charges three regions can be identified: near, intermediate and far field. In the near field the rapid cavity expansion originated by the blasting induces a shock wave. The propagation velocity of the wave front decreases with the distance from the source. In this region the grain skeleton is obliterated.

Soil compaction mainly takes place in the intermediate field. In this region the compression of the grain skeleton left behind by the shock wave leads to a rapid increase of the effective pressure followed by a delayed increase of pore water pressure due to the presence of gas bubbles. The loss of effective pressure (liquefaction), which starts a few minutes after the detonation, reaches a distance of about 20 m from the blasting point. The liquefaction leads to a collapse of the grain skeleton and hence to a compaction zone within the fill. Simultaneous measurements of earth and pore water pressures are necessary to detect liquefaction and determine its duration in the field.

No significant difference in compaction was observed between the "conventional" arrangement V1 and the arrangements V2 and V3. The level of vibration in the far field can be reduced by increasing the time delay between single detonations.

6 Acknowledgement

This study is based on a contribution to the research project "Reconstruction and Stabilisation of Dumps and Dump Slopes Endangered by Settlement Flow" supported by the LMBV mbH (Lusatian and Central German Mines Administration Company) and the German Federal Ministry of Education and Research (BMBF). The authors are indebted to W. Kuntze and F. Theil, LMBV, for thoughtful discussions and the support around the fieldtests. They are also grateful to Prof. G. Gudehus for his contributions and leading the project.

Field tests, especially their instrumentation, require a lot of preliminary tests and a careful installation and observation. Many members of the institute have contributed to the success of this part of the project, in particular S. Augustin, E. Bösinger, J. Benkler, N. Pralle, A. Schünemann, K. Wächter.

References

[1] CUDMANI, R., HUBER, G. (2000): *Evaluation of the state of a sandy open pit mine fill before and after blasting compaction by means of field testing.* This volume.

[2] GUDEHUS G., KUNTZE W., RAJU V.R., WARMBOLD U. (1996): *Sprengversuche zur Bodenverdichtung.* Tagungsband, Baugrundtagung Berlin. p. 523-536.

[3] NOVY, A., REICHEL, G., WARMBOLD, U., VOGT, A. (1999): *Geotechnische Untersuchungen und Verfahren bei der Sicherung setzungsfließgefährdeter Tagebaukippen der Niederlausitz.* BRAUNKOHLE/Surface Mining, **51**, 4, 465-478.

[4] OSINOV, V.A., CUDMANI, R. (2000): *Theoretical Investigation of the Cavity Expansion Problem based on a Hypoplasticity Model.* in preparation.

[5] PRALLE N., BAHNER, M. L. & BENKLER, J. (2000): *Computer tomographic analysis of indisturbed samples of loose sands reveal large gas inclusions in the phreatic zone.* (In preparation).

[6] RAJU, V. (1994): *Spontane Verflüssigung lockerer granularer Körper - Phänomene, Ursachen, Vermeidung.* Veröffentlichung des Institutes für Bodenmechanik und Felsmechanik der Universität Fridericiana in Karlsruhe, Heft 134.

[7] WARMBOLD, U. UND KUNTZE, W. (1994): *Sicherung böschungsnaher setzungsfließgefährdeter Kippenbereiche an Tagebau-Restseen.* Vorträge der Baugrundtagung, Köln, S. 331-348

T - #0628 - 101024 - C0 - 254/178/19 - PB - 9789058093189 - Gloss Lamination